THE 100+ SERIES™

Common Core Edition

ALGEBRA

Essential Practice for Advanced Math Topics

Carson-Dellosa Publishing, LLC
Greensboro, North Carolina

Visit carsondellosa.com for correlations to Common Core, state, national, and Canadian provincial standards.

Carson-Dellosa Publishing, LLC
PO Box 35665
Greensboro, NC 27425 USA
carsondellosa.com

ISBN 978-1-4838-0077-6

04-144167784

Table of Contents

Introduction

What are the Common Core State Standards for Middle School Mathematics?

In grades 6–8, the standards are a shared set of expectations for the development of mathematical understanding in the areas of ratios and proportional relationships, the number system, expressions and equations, functions, geometry, and statistics and probability. These rigorous standards encourage students to justify their thinking. They reflect the knowledge that is necessary for success in college and beyond.

Students who master the Common Core standards in mathematics as they advance in school will exhibit the following capabilities:

1. Make sense of problems and persevere in solving them.

Proficient students can explain the meaning of a problem and try different strategies to find a solution. Students check their answers and ask, "Does this make sense?"

2. Reason abstractly and quantitatively.

Proficient students are able to move back and forth smoothly between working with abstract symbols and thinking about real-world quantities that symbols represent.

3. Construct viable arguments and critique the reasoning of others.

Proficient students analyze problems by breaking them into stages and deciding whether each step is logical. They justify solutions using examples and solid arguments.

4. Model with mathematics.

Proficient students use diagrams, graphs, and formulas to model complex, real-world problems. They consider whether their results make sense and adjust their models as needed.

5. Use appropriate tools strategically.

Proficient students use tools such as models, protractors, and calculators appropriately. They use technological resources such as Web sites, software, and graphing calculators to explore and deepen their understanding of concepts.

6. Attend to precision.

Proficient students demonstrate clear and logical thinking. They choose appropriate units of measurement, use symbols correctly, and label graphs carefully. They calculate with accuracy and efficiency.

7. Look for and make use of structure.

Proficient students look closely to find patterns and structures. They can also step back to get the big picture. They think about complicated problems as single objects or break them into parts.

8. Look for and express regularity in repeated reasoning.

Proficient students notice when calculations are repeated and look for alternate methods and shortcuts. They maintain oversight of the process while attending to the details. They continually evaluate their results.

How to Use This Book

In this book, you will find a collection of 100+ reproducible practice pages to help students review, reinforce, and enhance Common Core mathematics skills. Use the chart provided on the next page to identify practice activities that meet the standards for learners at different levels of proficiency in your classroom.

Common Core State Standards* Alignment: Algebra

Domain: Ratios and Proportional Relationships	
Standard	**Aligned Practice Pages**
7.RP.A.2b	7
7.RP.A.2c	7
7.RP.A.3	8–10

Domain: Expressions and Equations	
Standard	**Aligned Practice Pages**
7.EE.A.1	26, 30–32, 52–57
7.EE.B.4b	85
8.EE.A.1	12–14, 24
8.EE.A.2	15, 16, 22
8.EE.A.3	11
8.EE.B.5	71–83
8.EE.C.7a	85, 86
8.EE.C.7b	27–29, 33–36, 38–49
8.EE.C.8a	51, 89, 90
8.EE.C.8b	91–94, 97

Domain: Functions	
Standard	**Aligned Practice Pages**
8.F.A.1	64–70, 80–83, 98
8.F.A.2	100

Domain: Geometry	
Standard	**Aligned Practice Pages**
8.G.B.7	20, 21

Domain: The Real Number System	
Standard	**Aligned Practice Pages**
7.NS.A.3	37
HSN-RN.A.1	14
HSN-RN.A.2	15, 16
HSN-RN.B.3	22

Domain: Seeing Structure in Expressions	
Standard	**Aligned Practice Pages**
HSA-SSE.A.2	53–62

Domain: Creating Equations	
Standard	**Aligned Practice Pages**
HSA-CED.A.1	45, 46, 59, 60, 62
HSA-CED.A.2	50, 51, 78–81

Domain: Reasoning with Equations and Inequalities	
Standard	**Aligned Practice Pages**
HSA-REI.A.1	17
HSA-REI.A.2	18, 19, 23
HSA-REI.B.3	38–49, 58–62, 85, 86
HSA-REI.C.5	95–97
HSA-REI.C.6	91–94
HSA-REI.D.10	64–71, 73, 74, 76, 77, 83, 100
HSA-REI.D.11	67–70, 72, 75, 78, 79, 82, 103
HSA-REI.D.12	87–90

Domain: Interpreting Functions	
Standard	**Aligned Practice Pages**
HSF-IF.A.1	100
HSF-IF.A.2	99
HSF-IF.C.7e	103

Domain: Building Functions	
Standard	**Aligned Practice Pages**
HSF-BF.B.4c	101, 102

Domain: Linear, Quadratic, and Exponential Models	
Standard	**Aligned Practice Pages**
HSF-LE.A.2	64, 69, 70, 79–81

Name________________________________

7.RP.A.2b, 7.RP.A.2c

Problem Solving: Work Problems

Anne can complete a project in 6 hours. It takes Will 9 hours to do the same job. How long will it take them if they work together?

Let x = number of hours for both to complete job.

$\frac{1}{6}$ = Anne's rate

$\frac{1}{9}$ = Will's rate

$$\frac{1}{6} \cdot x + \frac{1}{9} \cdot x = 1$$

$$\frac{x}{6} + \frac{x}{9} = 1$$

$$18 \cdot \frac{x}{6} + 18 \cdot \frac{x}{9} = 18 \cdot 1$$

$$3x + 2x = 18$$

$$5x = 18$$

$$x = \frac{18}{5} \text{ or } 3\frac{3}{5}$$

They can complete the project in $3\frac{3}{5}$ hours.

1. Bill can paint a closet in 2 hours. Bob can paint the same closet in 3 hours. How long will it take them to paint the closet working together?

2. Sally can address a box of envelopes in 30 minutes. Her brother Jim can address a box of envelopes in 1 hour. How long would it take both working together to address a box of envelopes?

3. Paul can mow the grass in 50 minutes, but it takes Dan three times as long. How long will it take them to mow the grass if they work together?

4. Using 1 drain, a swimming pool can be emptied in 45 minutes. Using a different drain, the job requires 1 hour and 15 minutes. How long will it take if both drains are opened?

5. Susan can sort the office mail in 15 minutes; but if Kathy helps, they can sort the mail in 8 minutes. How long would it take Kathy to sort the mail alone?

6. One pipe can fill a tank in 4 hours. A second pipe also requires 4 hours, but a third needs three hours. How long will it take to fill the tank if all three pipes are open?

Name________________________

7.RP.A.3

Problem Solving: Simple and Percent Problems

If $1,400 is added to an account earning 6% annually, the interest will amount to $192. How much was in the account originally?

Let x = original account

$$0.06\,(x + 1{,}400) = \$192$$
$$0.06 + 84 = 192$$
$$0.06x = 108$$
$$x = 1{,}800 \qquad \text{account had } \$1{,}800$$

1. How much simple interest can be earned in one year on $800 at 6%?

2. How long will it take $1,000 to double at 6% simple interest?

3. Sam invested $1,600, part at 5% and the rest at 6%. The money earned $85 in one year. How much was invested at 5%?

Hint:

	P x	r x	t =	I
Amount at 5%	x		1	
Amount at 6%	1,600 – x		1	

4. The Lewis family invested $900, part at 5% and the rest at 7%. The income from the investment was $58. How much was invested at 7%?

5. The Lockmores invested $7,000, part at 8% and part at $6\frac{1}{2}$%. If the annual return was $537.50, how much was invested at each rate?

6. BDLV Associates had $7,400 invested at $5\frac{1}{2}$%. After part of the money was withdrawn, $242 was earned on the remaining funds for one year. How much money was withdrawn?

7. Michael has $2,000 more invested at $8\frac{1}{2}$% than he does at $9\frac{3}{4}$%. If the annual return from each investment is the same, how much is invested at each rate?

8. Ms. Burke invested $53,650, part at 10.5% and the rest at 12%. If the income from the 10.5% investment is one third of that from the 12% investment, how much did she invest at each rate?

Name________________________________

7.RP.A.3

Problem Solving: Mixture Problems

How much water must be added to 20kg of a 10% salt solution to produce a 5% solution?

Let x = amount of water

$$10\% \cdot 20 + 0\%x = 5\% \cdot (x + 20)$$
$$2 + 0 = 0.05x + 1$$
$$1 = 0.05x$$
$$20 = x$$

20kg of water

1. How much water must be added to 60kg of an 80% acid solution to produce a 50% solution?

2. How much water must be evaporated from 8 grams of a 30% antiseptic solution to produce a 40% solution?

3. How many grams of alcohol must be added to 40 grams of a 15% alcohol solution to obtain a 20% alcohol solution?

4. How many quarts of antifreeze must be added to 15 quarts of a 30% antifreeze solution to obtain a 50% antifreeze solution?

5. A candy mixture is created with 2 types of candy, one costing \$4 per pound and the other \$3.50 per pound. How much of each type is needed for a 5 pound box that costs \$18?

6. A seed company mixes two types of seed for bird feeding. One costs \$1.10 per kg and the other costs \$2.25 per kg. How much of each type of seed is needed to produce 6kg at a cost of \$8.90?

7. A farmer wants to mix milk containing 6% butterfat with 2 quarts of cream that is 15% butterfat to obtain a mixture that is 12% butterfat. How much milk containing 6% butterfat must he use?

8. A store owner has 12 pounds of pasta worth 70¢ a pound. She wants to mix it with pasta worth 45¢ a pound so that the total mixture can be sold for 55¢ a pound (without any gain or loss). How much of the 45¢ pasta must she use?

Name________________________________

7.RP.A.3

Calculating Compound Interest

Compound Interest

$$A = P\left(1 + \frac{r}{n}\right)^{nt}$$

where A = amount, P = principal, r = rate, t = time in years, and n = number of times compounded per year.

Solve the story problems assuming no deposits or withdrawals.

1. Heather received $100 for her 13th birthday. If she saves it in a bank with 3% interest compounded quarterly, how much money will she have in the bank by her 16th birthday?

2. Roland earned $1,500 last summer. If he deposited the money in a certificate of deposit that earns 4% interest compounded monthly, how much money will he have next summer?

3. The C.R.E.A.M. Company has an employee savings plan. If an employee makes an initial contribution of $2,500 and the company pays 5% interest compounded quarterly, how much money will the employee have after 10 years?

4. Juan invests $7,500 at 6% interest for one year. How much money would he have if the interest were compounded
   a. Yearly?
   b. Daily?
   c. Why are the amounts in answers *a* and *b* different?

5. Carmen is saving for a new car that costs $15,000. If she puts $5,000 in an account that earns 6% interest compounded monthly, how long will it take for her to save enough money to buy the car?

Name____

8.EE.A.3

Scientifically Speaking

Scientific notation is used to write very large and very small numbers. A number in scientific notation is the product of a number between 1 and 10 and a power of 10.

Examples: $45{,}000{,}000 = 4.5 \times 10^7$ $0.00000625 = 6.25 \times 10^{-6}$

Write each measurement in scientific notation. Then, write the problem letter above the value of the exponent to complete the statement at the bottom of the page.

C The population of China is greater than 1,250,000,000.

E Scientists at Oak Ridge National Laboratory have sent an electric current of 2,000,000 amperes/cm^2 down a wire.

E The diameter of an electron is 0.0000000000011 cm.

E In an election in India, more than 343,350,000 people voted.

E The Earth's mass is 5,980,000,000,000,000,000,000 metric tons.

G The Greenland–Canada boundary is about 1,700 miles long.

I The isotope lithium 5 decays in 0.0000000000000000000044 seconds.

I The isotope tellurium 128 has a half-life of 1,500,000,000,000,000,000,000,000 years.

M A microbe strain of H39 has a diameter of 0.0000003 m.

N A nugget of platinum found in 1843 weighed 340 ounces.

N In 1996 the United States national debt was $5,129,000,000,000.

O A Saudi Arabia oil field contains about 82,000,000,000 barrels.

P The fastest planet, Mercury, travels at 107,000 mph.

R Sales of the record "White Christmas" exceeded 30,000,000.

V In 1973, a vulture flying at 37,000 ft. collided with an aircraft.

X The wavelength of an X ray is about 0.0000000015 m.

What the decimal point said about scientific notation:

"It's a ___ ___ ___ ___ ___ ___   ___ ___ ___ ___ ___ ___ ___ ___ ___ ___ !"

−7 10 4 −22 2 3   6 −9 5 −12 7 24 8 12 9 21

Name________________________________

8.EE.A.1

Manipulating Powers

1) $(a^x)^y = a^{xy}$

2) $a^x \cdot a^y = a^{x+y}$

3) $\frac{a^x}{a^y} = a^{x-y}$

4) $(ab)^x = a^x b^x$

5) $\left(\frac{a}{b}\right)^x = \frac{a^x}{b^x}$

6) $a^{-x} = \frac{1}{a^x}$

7) $\frac{1}{a^{-x}} = a^x$

Simplify each expression.

Example: $(x^2)^4 = x^{2 \cdot 4} = x^8$

1. $x^4 \cdot x^2$

2. $\frac{x^8}{x^6}$

3. $(x^2y)^3$

4. $\left(\frac{x}{y^3}\right)^5$

5. y^{-15}

6. $\frac{1}{x^{-15}}$

7. $\frac{a^6}{a^9}$

8. $(2c^2)^3$

9. $\frac{n^4 \cdot n^6}{n^8 \cdot n^2}$

10. $4a^5 \cdot 3a^3$

11. $\left(\frac{v}{3}\right)^4 \cdot \left(\frac{5}{v}\right)^2$

12. $(x^{-2})^2$

13. $\left(\frac{2}{x}\right)^{-1}$

Name____________________________________ 8.EE.A.1

Manipulating Powers

14. $(x^{-2} \cdot y)^{-3}$

15. $\dfrac{12x^5}{3x^7}$

16. $\dfrac{8d}{(10d^{-4})(9d^2)}$

17. $^{-}2x^{-2}$

18. $x^{\frac{1}{3}} \cdot x^{\frac{2}{3}}$

19. $\left(\dfrac{8x}{125}\right)^{-2}$

20. $\dfrac{a^4 \cdot b^6 \cdot a^9}{b^{-2}}$

21. $\dfrac{x^{-4}y^{-6}}{x^2y^5z}$

22. $\left(\dfrac{x^2}{(xz)^2}\right)^{-2}$

23. $\left(\dfrac{x^2y^1z}{a^4b^{-7}}\right)^{-3}$

24. $(x^2y^2)^{-2} \cdot x^4y^{19}$

25. $\left(\dfrac{x^{-4}}{y^6}\right)^3 \cdot \left(\dfrac{x}{y^{-4}}\right)^{-4}$

26. $(a^2b^1c^8)^6 \cdot a^{-9} \cdot b^4 \cdot x$

27. $\left(\dfrac{x^{-4}b^{-1}}{4}\right)^{-3} \cdot 2x^5$

28. $(a^9b^{-2}c^1)^{-4} \cdot \left(\dfrac{ab}{x}\right)^3$

29. $\left(\dfrac{x^{-4}y^{-6}z^{10}}{a^1b^2c^{-4}}\right)^{-2} \cdot \left(\dfrac{a^1bc^{-4}}{x^6yz^9}\right)$

Name________________________________ 8.EE.A.1, HSN-RN.A.1

Evaluating Rational Exponents

Simplify the expressions.

Example 1: $\left(\frac{64}{343}\right)^{\frac{1}{3}} = \frac{64^{\frac{1}{3}}}{343^{\frac{1}{3}}} = \frac{4}{7}$

Example 2: $\left(\frac{1000}{64}\right)^{-\frac{2}{3}} = \left(\frac{64}{1000}\right)^{\frac{2}{3}} = \left(\frac{(64)^{\frac{1}{3}}}{1000^{\frac{1}{3}}}\right)^{2} = \left(\frac{4}{10}\right)^{2} = \frac{16}{100} = \frac{4}{25}$

1. $\left(\frac{216}{729}\right)^{\frac{2}{3}}$

2. $\left(\frac{81}{16}\right)^{\frac{3}{2}}$

3. $\left(\frac{1}{32}\right)^{\frac{3}{5}}$

4. $\left(\frac{1}{25}\right)^{\frac{1}{2}}$

5. $\left(\frac{256}{625}\right)^{\frac{3}{4}}$

6. $\left(\frac{81}{256}\right)^{\frac{-3}{4}}$

7. $\left(\frac{121}{36}\right)^{\frac{-1}{2}}$

8. $\left(\frac{32}{243}\right)^{\frac{2}{5}}$

9. $\left(\frac{729}{343}\right)^{\frac{-2}{3}}$

10. $\left(\frac{4}{81}\right)^{\frac{-3}{2}}$

11. $\left(\frac{343}{64}\right)^{\frac{2}{3}}$

Calculator Challenge:

12. $\left(\frac{225}{289}\right)^{\frac{-5}{2}}$

As a check, for each problem number, substitute the answer.

$1 \cdot 2 \cdot 3 \div 4 \cdot 5 \cdot 6 \div 7 \div 8 \div 9 \div 10 \cdot 11 = \frac{22}{25}$

__ • __ • __ ÷ __ • __ • __ ÷ __ ÷ __ ÷ __ ÷ __ • __ = $\frac{22}{25}$

Name________________________________

8.EE.A.2, HSN-RN.A.2

Simplifying Radicals

All variables are non-negative numbers.

$$\sqrt{18x^3y^2} = \sqrt{9 \cdot 2\, x^2 \cdot x \cdot y^2}$$
$$= 3xy\,\sqrt{2x}$$

1. $\sqrt{100} =$

2. $\sqrt{75} =$

3. $-\sqrt{144a^2} =$

4. $\sqrt{128x^3} =$

5. $2\sqrt{1000} =$

6. $\sqrt{15a^8b} =$

7. $\sqrt{16c^2d^2} =$

8. $2\sqrt{27x^5y} =$

9. $-\sqrt{20xy^2} =$

10. $\sqrt{50a^3} =$

11. $\sqrt{96bc^2d^5} =$

12. $^{-}3\sqrt{150a^7c^2} =$

13. $\sqrt{27a^2} =$

14. $2\sqrt{50x^2yz^3} =$

15. $\sqrt{243m^5n^2} =$

16. $-\sqrt{320y^9z^{10}} =$

Name________________________________ 8.EE.A.2, HSN-RN.A.2

Radicals Challenge

Simplify each radical.

Example: $\sqrt[3]{24m^3x^5}$

2 12 $\quad 2 \cdot m \cdot x\sqrt[3]{3x^2}$

2 6 $\quad 2mx\sqrt[3]{3x^2}$

2 3

1. $\sqrt{49m^2t^3}$

2. $\sqrt[4]{81x^4}$

3. $\sqrt{64a^2b^4}$

4. $\sqrt[4]{16x^5y^3}$

5. $\sqrt[3]{27x^6y^9}$

6. $\sqrt[10]{1000x^{12}y^{100}}$

7. $\sqrt[3]{343a^6}$

8. $-\sqrt{(2x+1)^2}$

9. $\sqrt[3]{(x+1)^6}$

10. $\sqrt{x^2+2x+1}$ **Hint:** factor

11. $\sqrt{4x^2-12x+9}$

Name________________________________ HSA-REI.A.1

Products of Radicals Quiz

Randy did not do well on the Radicals Quiz. Find and correct his seven errors.

Products of Radicals Quiz

Name *Randy*

Quiz Hints: All variables are non-negative numbers.

Multiply radicals and simplify.

Examples: $4\sqrt{3} \cdot 2\sqrt{18} \Rightarrow 8\sqrt{54} \Rightarrow 8\sqrt{9 \cdot 6} \Rightarrow 8 \cdot 3\sqrt{6} = 24\sqrt{6}$

$\sqrt{2a} \cdot \sqrt{6a} \Rightarrow \sqrt{12a^2} \Rightarrow \sqrt{4 \cdot 3 \cdot a^2} = 2a\sqrt{3}$

1. $\sqrt{2} \cdot \sqrt{8} = 4$

2. $5\sqrt{5} \cdot 3\sqrt{14} = 15\sqrt{70}$

3. $\sqrt{5b} \cdot \sqrt{10b} = 2b\sqrt{5}$

4. $a\sqrt{2x} \cdot x\sqrt{6x} = 2ax\sqrt{3x}$

5. $2m\sqrt{7mn} \cdot 3\sqrt{7m} = 42m^2\sqrt{n}$

6. $^-5a\sqrt{2a^4b} \cdot 4b\sqrt{12a^3b^4}$

   $= {}^-40ab\sqrt{6a^7b}$

7. $2\sqrt{5}\,(-\sqrt{3x}) = {}^-2\sqrt{15x}$

8. $5\sqrt{6} \cdot 2\sqrt{2} = 30\sqrt{2}$

9. $\sqrt{x} \cdot \sqrt{9x} = 3x$

10. $\sqrt{2x} \cdot \sqrt{10x^2y} = 5y\sqrt{2xy}$

11. $4x\sqrt{5} \cdot \sqrt{8xy^2} = 8xy\sqrt{10x}$

12. $2\sqrt{x^3} \cdot 4\sqrt{x} = 8x^4$

13. $^-7\sqrt{3y} \cdot \sqrt{6y} = {}^-14y\sqrt{3}$

14. $\sqrt{xy} \cdot \sqrt{xy} = \sqrt{xy}$

15. $\sqrt{x^3y^5} \cdot \sqrt{x^2y} = x^2y^3\sqrt{x}$

Name________________________________

HSA-REI.A.2

Quotients of Radicals

Quotients — Rationalizing the denominator.

$$\sqrt{\frac{7}{8}} = \frac{\sqrt{7}}{\sqrt{8}} = \frac{\sqrt{2}}{\sqrt{2}} = \frac{\sqrt{14}}{\sqrt{16}} = \frac{\sqrt{14}}{4}$$

$$\sqrt{\frac{2a^4b^3}{27x^3}} = \frac{\sqrt{2a^4b^3}}{\sqrt{27x^3}} \cdot \frac{\sqrt{3x}}{\sqrt{3x}} = \frac{\sqrt{6a^4b^3x}}{\sqrt{81x^4}} = \frac{a^2b\sqrt{6bx}}{9x^2}$$

1. $\sqrt{\frac{2ab^2}{c^2d}} =$

2. $\sqrt{\frac{2x}{3y}} =$

3. $\sqrt{\frac{19x^2}{32}} =$

4. $\sqrt{\frac{4a^2b}{x^8y^7}} =$

5. $x\sqrt{\frac{5d}{3x^2}} =$

6. $\sqrt{\frac{7a^2}{8cd}} =$

7. $\sqrt{\frac{n^2}{7}} =$

8. $\sqrt{\frac{8}{25}} =$

9. $\sqrt{\frac{3\sqrt{2}}{\sqrt{3}}} =$

10. $\sqrt{\frac{4x^2}{25}} =$

11. $\sqrt{\frac{11y^3}{9}} =$

12. $\sqrt{\frac{25}{3x}} =$

13. $\sqrt{\frac{3}{6x^3}} =$

14. $\frac{\sqrt{8x^2y}}{\sqrt{2y}} =$

Name________________________________ HSA-REI.A.2

Sums and Differences of Radicals

$$3\sqrt{9xy^4} - y\sqrt{16xy^2} + 2y^2\sqrt{25x} = 9y^2\sqrt{x} - 4y^2\sqrt{x} + 10y^2\sqrt{x} = 15y^2\sqrt{x}$$

$$10\sqrt{\frac{1}{5}} + 4\sqrt{18} + 3\sqrt{45} - 8\sqrt{\frac{1}{2}} = 2\sqrt{5} + 12\sqrt{2} + 9\sqrt{5} - 4\sqrt{2} = 11\sqrt{5} + 8\sqrt{2}$$

1. $3\sqrt{7} - 4\sqrt{7} + 2\sqrt{7} =$

2. $4\sqrt{27} - 2\sqrt{48} + \sqrt{147} =$

3. $5\sqrt{3} - 4\sqrt{7} - 3\sqrt{3} + \sqrt{7} =$

4. $5\sqrt{x} - 3\sqrt{x} + a\sqrt{x} =$

5. $4\sqrt{\frac{1}{2}} + 2\sqrt{18} - 6\sqrt{\frac{2}{9}} =$

6. $\sqrt{63} - \sqrt{28} - \sqrt{7} =$

7. $6\sqrt{3} - 2\sqrt{75} + 4\sqrt{\frac{3}{16}} =$

8. $\sqrt{50} + \sqrt{98} - \sqrt{75} + \sqrt{27} =$

9. $2x\sqrt{ab} - 2y\sqrt{ab} + 4x\sqrt{ab} =$

10. $2b\sqrt{3c} + b\sqrt{5c} + b\sqrt{3c} - 2b\sqrt{5c} =$

11. $4\sqrt{c^3d^3} + 3cd\sqrt{4cd} - 2c\sqrt{9cd^3} =$

12. $8\sqrt{12} - \sqrt{10}\sqrt{\frac{1}{5}} - 108 + \sqrt{125} =$

13. $x\sqrt{4x} + \sqrt{x^3} =$

14. $3x\sqrt{7} + \sqrt{28x^2} - \sqrt{63x^2} =$

Name________________________________

Pythagorean Magic

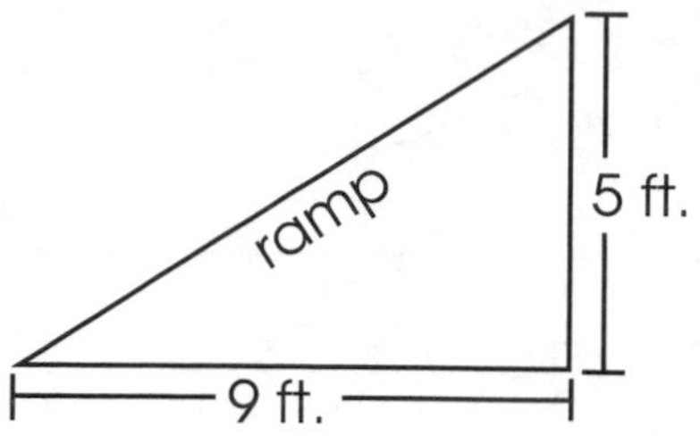

How long is the ramp?

$9^2 + 5^2 = \text{ramp}^2$

ramp = 10.3 feet

1. How high is the flag pole?

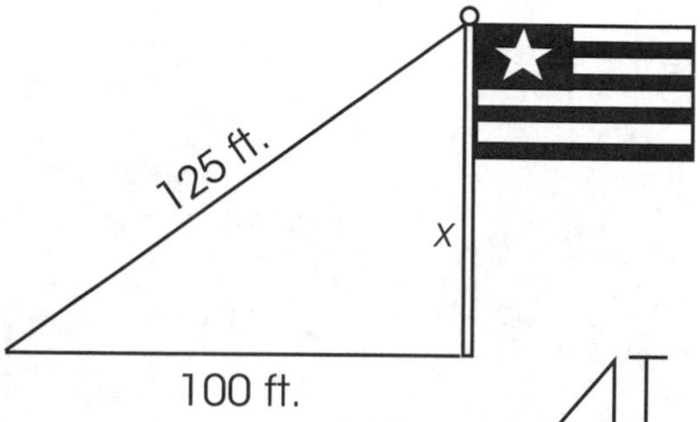

2. How long is the longest side of the sail?

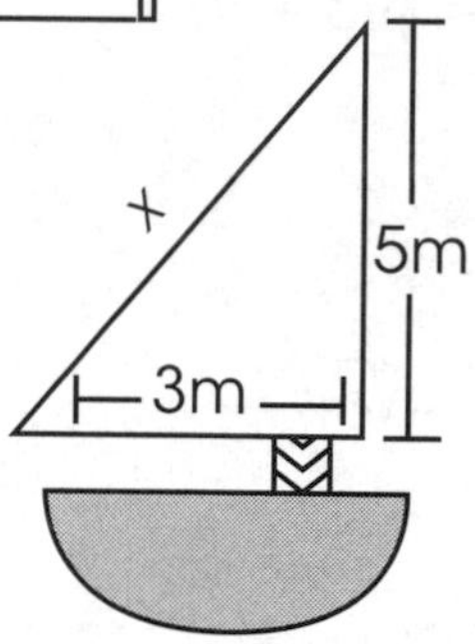

3. A 10-foot ladder is leaning against a house with its base 4 feet from the base of the house. How far up the house does the ladder touch the house? (**Hint:** Draw a picture.)

4. A 5-foot tall tree casts an 8-foot shadow on the ground. How far is it from the end of the shadow to the top of the tree? (**Hint:** Draw a picture.)

5. A guy wire is secured into the ground 15 feet from the base of a 36-foot pole. How long is the guy wire if it is attached at the top of the 36-foot pole? (**Hint:** Draw a picture.)

6. An airplane travels due east 65 miles, and then due north 72 miles. How far is the airplane from its starting point? (**Hint:** Draw a picture.)

Manipulating Special Right Triangles

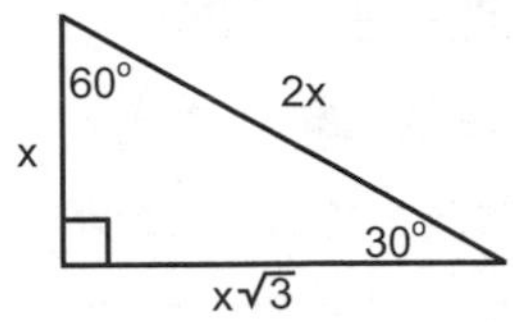

30° 60° 90° Right Triangle

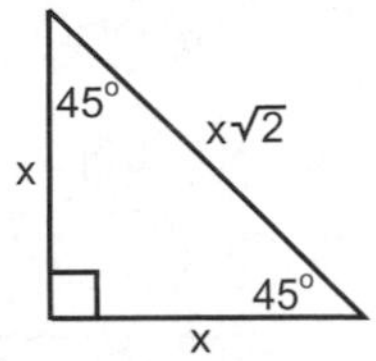

45° 45° 90° Right Triangle

Example: Find the missing sides of each triangle. Leave in radical form.
Note: Since these are right triangles, you can check your answer using the Pythagorean theorem.

1.

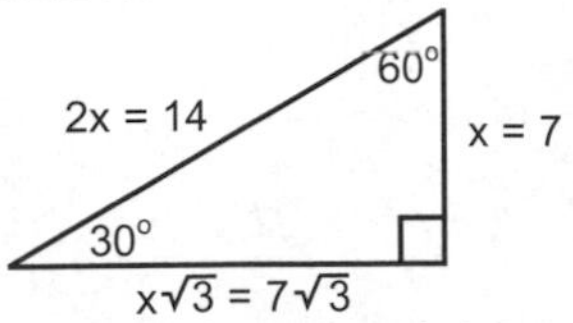

2.

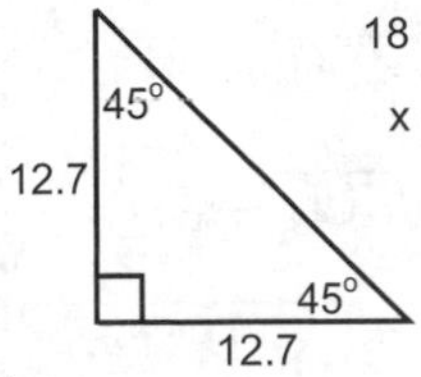

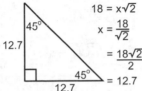

$18 = x\sqrt{2}$

$x = \frac{18}{\sqrt{2}}$

$= \frac{18\sqrt{2}}{2}$

$= 12.7$

Find the missing sides of each triangle. Check your answers using the Pythagorean theorem.

1.

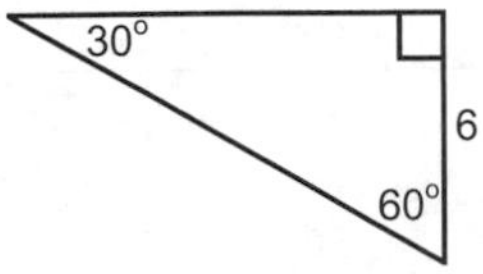

2.

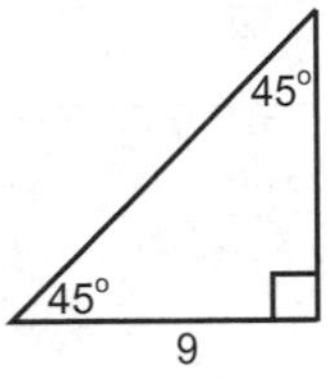

3.

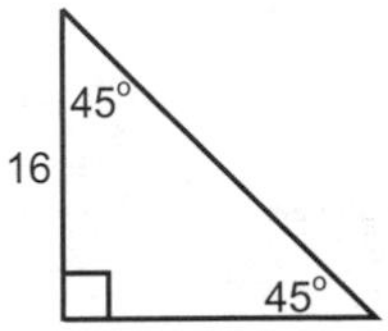

4.

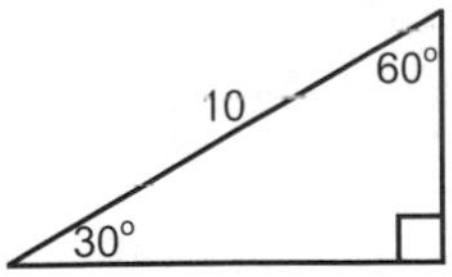

5.

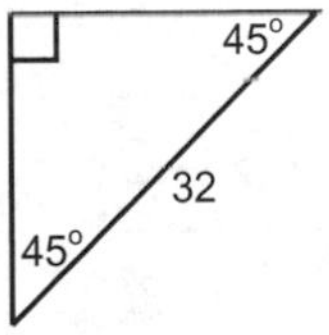

6.

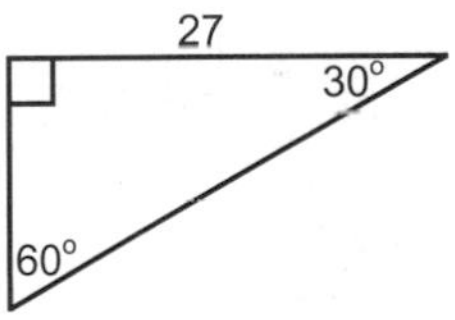

7.

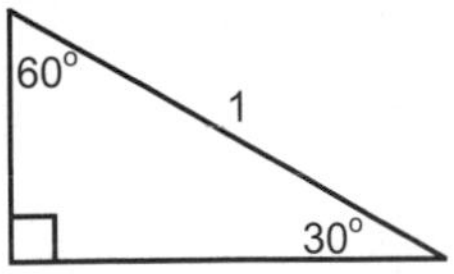

8.

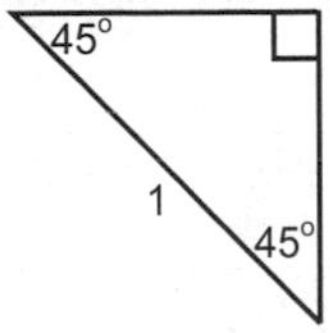

Cross out the correct answers below. Use the remaining letters to complete the statement.

9 O	9/2 I	16√2 N	1/4 S	24 H	9√2 F	25 A	√2 L	16√2 O	√3 F	3 T	12 U
10 H	5 R	81 E	1/2 T	6√3 H	9√3 O	5/2 H	1/8 Y	7 P	16√2 F	7√2 O	√2/2 T
5√3 H	√2/2 E	11√2 T	11/2 E	15 N	25/2 U	16 L	√3/2 E	3/2 S	16√3 E	18√3 G	15/2 .

In a 30–60 degrees right triangle, the side opposite the 30-degree angle

Name________________________________

8.EE.A.2, HSN-RN.B.3

Radical Animals

B is a square root of A if $A = B^2$.
5 and $^-5$ are square roots of 25 since 5^2 and $^-5^2 = 25$.
A radical $\sqrt{}$ sign implies the positive root: $\sqrt{25} = 5$.

Solve the problems to determine the oldest recorded ages for the given animals.

1. Black widow spider $\sqrt{5{,}000} \cdot \sqrt{2} =$ _________ days
2. Bedbug $\sqrt{10{,}000} + \sqrt{6{,}400} + \sqrt{4} =$ _________ days
3. Common housefly $\sqrt{400} - \sqrt{9} =$ _________ days
4. Queen ant $\sqrt{\frac{1}{25}} \cdot \sqrt{8{,}100} =$ _________ years
5. Giant centipede $\sqrt{36} + \sqrt{16} =$ _________ years
6. Goldfish $\sqrt{2{,}500} - \sqrt{81} =$ _________ years
7. Common toad $\sqrt{6{,}400} \div \sqrt{4} =$ _________ years
8. Boa constrictor $\sqrt{4} \cdot \sqrt{400} =$ _________ years
9. American alligator $\sqrt{36} \cdot \sqrt{121} =$ _________ years
10. Blue whale $\sqrt{2{,}500} - \sqrt{25} =$ _________ years
11. Octopus $\sqrt{\frac{1}{16}} \cdot \sqrt{144} =$ _________ years
12. Asian elephant $\sqrt{6{,}400} + \sqrt{1} =$ _________ years
13. Andean condor $\sqrt{144} \cdot \sqrt{36} =$ _________ years
14. Monarch butterfly $\sqrt{1.21} =$ _________ years
15. Giant salamander $10\sqrt{25} + \sqrt{1} =$ _________ years
16. Hedgehog $\sqrt{144} + \sqrt{4} =$ _________ years
17. Horse $2\sqrt{900} + \sqrt{4} =$ _________ years
18. Ostrich $5\sqrt{16} + 5\sqrt{100} - \sqrt{64} =$ _________ years
19. Sea anemone $\sqrt{25} \cdot 2\sqrt{81} =$ _________ years
20. Tortoise $\sqrt{22{,}500} + \sqrt{4} =$ _________ years

Name________________________________

HSA-REI.A.2

Solving Radical Equations

$3 + \sqrt{x} = 6$ $\sqrt{x} = \sqrt{3}$ $(\sqrt{x})^2 = (3)^2$ $x = 9$

1. $\sqrt{x - 1} = 4$

2. $4 = 5\sqrt{x}$

3. $\sqrt{x + 3} = 1$

4. $8 = \sqrt{5a + 1}$

5. $2\sqrt{x} = 5$

6. $\sqrt{7 + 3x} = 4$

7. $\sqrt{4 - x} = 7$

8. $4 + \sqrt{x + 1} = 5$

9. $\dfrac{\sqrt{5 - 2x}}{3} = 1$

10. $\sqrt{4x - 3} = \sqrt{x}$

11. $5 = \dfrac{15}{\sqrt{2a - 3}}$

12. $6 - \sqrt{y - 5} = 3$

13. $2\sqrt{5} = 3\sqrt{x}$

14. $2\sqrt{x} = 4\sqrt{3}$

Name____

8.EE.A.1

Exponential Decay (Half-Life)

$y = a\left(\frac{1}{2}\right)^x$ where a = initial amount

x = number of half-lives = $\frac{\text{time}}{\text{half-life}}$

y = remaining

Solve each problem.

1. There are 10 grams of Curium-245 which has a half-life of 9,300 years. How many grams will remain after 37,200 years?

2. There are 80 grams of Cobalt-58 which has a half-life of 71 days. How many grams will remain after 213 days?

3. The half-life of Rhodium-105 is 1.5 days. If there are initially 7500 grams of this isotope, how many grams would remain after 30 days?

4. Two hundred ten years ago there were 132,000 grams of Cesium-137. How much is there today? The half-life of Cesium is 30 years.

5. In a nuclear reaction, 150 grams of Plutonium-239 are produced. How many grams would remain after one million years? The half-life of Plutonium-239 is 24,400 years.

6. Using carbon dating, scientists can determine how old a fossil is by how much Carbon-14 is present. If an average animal carcass contains 1 gram of Carbon-14, how old is a fossil with 0.0625 grams of Carbon-14? The half-life of Carbon-14 is 5730 years.

Name________________________________

Just for Fun

Make your own matrix.

One week there is a birthday party every day. No two children are invited to the same party. Find out the day that each child attends a party. Start your matrix with Sunday and continue through Saturday.

1. Lisa and Pat don't go to a party on Friday or Saturday.
2. Pat and Alice don't go on a Tuesday, but Sandy does.
3. Jennifer goes to a party on a Wednesday.
4. Jim goes to a party the day after Jennifer.
5. Lisa goes to a party the day before Pat.
6. Paul goes to a party on a Saturday.

Name________________________________ 7.EE.A.1

Simplifying Fractions

$$\frac{5x^2 + 30x - 35}{5 - 5x^2} = \frac{\cancel{5}(x+7)\cancel{(x-1)}}{\cancel{-5}(x+1)\cancel{(x-1)}} = -\frac{x+7}{x+1}$$

1. $\frac{8a - 8b}{a^2 - b^2} =$

2. $\frac{x^2 + 8x + 16}{x^2 - 16} =$

3. $\frac{12 - 4a}{a^2 + a - 12} =$

4. $\frac{t^2 + 4t - 5}{t^2 + 9t + 20} =$

5. $\frac{z^2 - 4z - 5}{z^2 + 4z - 45} =$

6. $\frac{6b^3 - 24b^2}{b^2 + b - 20} =$

7. $\frac{-x^2 + 8x - 12}{x - 2} =$

8. $\frac{2a^3 + a^2 - 3a}{6a^3 + 5a^2 - 6a} =$

9. $\frac{x^2 - 9}{x^2 + x - 6} =$

10. $\frac{3x^2 + 2x - 1}{x^2 + 3x + 2} =$

11. $\frac{x^2 + 5x}{x^2 - 25} =$

12. $\frac{a^2 - 11a + 30}{a^2 - 9a + 18} =$

13. $\frac{2y^3 - 12y^2 + 2y}{y^2 - 6y + 1} =$

14. $\frac{a + b}{a^2 + 2ab + b^2} =$

Name________________________________ 8.EE.C.7b

Multiplying Fractions

$$\frac{^-22cd^2}{2d} \cdot \frac{17c^2d}{17d} = \frac{^-1 \cdot \cancel{2}2 \cdot 11 \cdot \overset{d}{\cancel{cd^2}}}{\cancel{2d}} \cdot \frac{\cancel{17}c^2\cancel{d}}{\cancel{17d}} = {}^-11c^3d$$

$$\frac{3x-6}{6x+6} \cdot \frac{x^2+3x+2}{x^2-3x+2} = \frac{\cancel{3}\cancel{(x-2)}}{\underset{2}{\cancel{6}}\cancel{(x+1)}} \cdot \frac{(x+2)\cancel{(x+1)}}{\cancel{(x-2)}(x-1)} = \frac{x+2}{2(x-1)}$$

1. $\frac{24r^2s^2}{3s} \cdot \frac{^-21s}{r} =$

2. $\frac{x^2y}{z^2} \cdot \frac{z}{xy} =$

3. $\frac{2t+16}{4t} \cdot \frac{10t^2}{3t+24} =$

4. $\frac{x^2-1}{x} \cdot \frac{x^2}{x-1} =$

5. $\frac{a+b}{a-b} \cdot \frac{a^2-b^2}{a+b} =$

6. $\frac{a^2-4}{a^2-1} \cdot \frac{a-1}{a-1} =$

7. $\frac{2x+2}{x-1} \cdot \frac{x^2+x-2}{x^2-x-2} =$

8. $\frac{z^2-6z-7}{z^2+z} \cdot \frac{z^2-z}{3z-21} =$

9. $\frac{c^2-6c-16}{c^2+4c-21} \cdot \frac{c^2-8c+15}{c^2+9c+14} =$

10. $\frac{x+8}{x^2-x-12} \cdot \frac{x^2-6x+8}{x^2+6x-16} =$

11. $\frac{h^2-2h-3}{h^2-9} \cdot \frac{h^2+5h+6}{h^2-1} =$

12. $\frac{x^2-y^2}{x^2+4xy+3y^2} \cdot \frac{x^2+xy-6y^2}{x^2+xy-2y^2} =$

13. $\frac{30+y-y^2}{25-y^2} \cdot \frac{y^2}{y^2-6y} \cdot \frac{y^2-y-12}{y^2-9} =$

14. $\frac{5m+5n}{m^2-n^2} \cdot \frac{m^2-mn}{(m+n)^2} =$

Name________________________________

8.EE.C.7b

Dividing Fractions

$$\frac{12a^2b^2}{21xy^2} \div \frac{4ab^2}{7y^2} = \frac{12a^2b^2}{21xy^2} \cdot \frac{7y^2}{4ab^2} = \frac{a}{x}$$

1. $\frac{b+2}{b^2-9} \div \frac{1}{b-3} =$

2. $\frac{c^2+2cd}{2cd+d^2} \div \frac{c^3+2c^2d}{cd+d^2} =$

3. $\frac{x^2+3x^3}{4-x^2} \div \frac{x+4x^2+3x^3}{2x+x^2} =$

4. $\frac{a^2-a-20}{a^2+7a+12} \div \frac{a^2-7a+10}{a^2+9a+18} =$

5. $\frac{6a^2-a-2}{12a^2+5a-2} \div \frac{4a^2-1}{8a^2-6a+1} =$

6. $\frac{a^3-6a^2+8a}{5a} \div \frac{2a-4}{10a-40} =$

7. $\frac{12x+36}{x^2-2x-8} \div \frac{15x+45}{x^2+x-20} =$

8. $\frac{x^2-y^2}{x^2+2xy+y^2} \div \frac{x-y}{x+y} =$

9. $(y^2-9) \div \frac{y^2+8y+15}{2y+10} =$

10. $\frac{x^2-4x+4}{3x-6} \div (x-2) =$

11. $\frac{(2a)^3}{(4bc)^3} \div \frac{16a^2}{8b^2c^3} =$

12. $\dfrac{\frac{26c^2}{5c^2d}}{\frac{13c^3}{25d^3}} =$

Name________________________________ 8.EE.C.7b

Combination Problems

Express answers in simplest form.

1. $\frac{6x}{3x - 7} \cdot \frac{9x - 21}{21} \div \frac{x^2}{35} =$

2. $\frac{x^2 - x - 6}{x^2 + 2x - 15} \cdot \frac{x^2 - 25}{x^2 - 4x - 5} \div \frac{x^2 + 5x + 6}{x^2 - 1} =$

3. $\frac{x - y}{x + y} \div \frac{5x^2 - 5y^2}{3x - 3y} \cdot \frac{(x + y)^2}{x^2 - y^2} =$

4. $(b^2 - 9) \div \frac{b^2 + 8b + 15}{2b + 10} \div (b - 3) =$

5. $\frac{a^3b^3}{a^3 - ab^2} \div \frac{abc}{a - b} \cdot \frac{ab + bc}{ab} =$

6. $\frac{x^2 + 16x + 64}{x^2 - 9} \div \frac{x^2 - 64}{x + 3} \cdot (x^2 - 11x + 24) =$

Name________________________________ 7.EE.A.1

Adding and Subtracting with Like Denominators

$$\frac{5}{17} + \frac{3}{17} - \frac{11}{17} = \frac{5 + 3 - 11}{17} = \frac{^-3}{17}$$

$$\frac{5a + 3c}{2a} - \frac{a - c}{2a}$$
$$= \frac{5a + 3c - (a - c)}{2a}$$
$$= \frac{5a + 3c - a + c}{2a}$$
$$= \frac{4a + 4c}{2a}$$
$$= \frac{\not{2} \cdot 2(a + c)}{\not{2}a}$$
$$= \frac{2(a + c)}{a}$$

1. $\frac{2}{x} - \frac{8}{x} + \frac{3}{x} =$

2. $\frac{3a}{5b} + \frac{2a}{5b} =$

3. $\frac{r}{6} - \frac{5t}{6} =$

4. $\frac{x + y}{2} - \frac{x}{2} =$

5. $\frac{c}{c - d} - \frac{d}{c - d} =$

6. $\frac{6a}{a + d} + \frac{6d}{a + d} =$

7. $\frac{x^2}{x - 2} - \frac{4}{x - 2} =$

8. $\frac{c^2}{c^2 - 4} - \frac{6c + 16}{c^2 - 4} =$

9. $\frac{x^2 - 7x}{(x - 3)^2} + \frac{12}{(x - 3)^2} =$

10. $\frac{x^2}{2x + 14} - \frac{49}{2x + 14} =$

11. $\frac{7x}{2y + 5} - \frac{6x}{2y + 5} =$

12. $\frac{y + 4}{y - 5} - \frac{3y + 1}{y - 5} =$

13. $\frac{2x - 3}{2} - \frac{6x - 5}{2} =$

14. $\frac{8a - 1}{5} - \frac{3a - 6}{5} =$

Name________________________________ 7.EE.A.1

Adding and Subtracting with Unlike Denominators

$$\frac{1}{7} - \frac{a}{b} = \frac{1 \cdot b}{7 \cdot b} - \frac{7 \cdot a}{7 \cdot b}$$
$$= \frac{b}{7b} - \frac{7a}{7b}$$
$$= \frac{b - 7a}{7b}$$

$$\frac{3}{x^2} + \frac{5}{2xy} - \frac{4}{3y^2} = \frac{3 \cdot 6y^2}{x^2 \cdot 6y^2} + \frac{5 \cdot 3xy}{2xy \cdot 3xy} - \frac{4 \cdot 2x^2}{3y^2 \cdot 2x^2}$$
$$= \frac{18y^2 + 15xy - 8x^2}{6x^2y^2}$$

1. $\frac{1}{x} + \frac{1}{y} =$

2. $\frac{3n}{7} + \frac{n}{14} =$

3. $\frac{2x}{3} + \frac{5y}{2} =$

4. $\frac{x}{3} + \frac{x^2}{5} =$

5. $\frac{2x}{x^2y} - \frac{y}{xy^2} =$

6. $\frac{5}{12xy} + \frac{3}{4x} =$

7. $\frac{a}{b} - \frac{c}{d} =$

8. $\frac{8}{x} + \frac{3}{xy} =$

9. $\frac{4x - 1}{3x} + \frac{x - 8}{5x} =$

10. $\frac{2x + 1}{4} - \frac{x - 1}{8} =$

11. $\frac{a + 2b}{3} + \frac{a + b}{2} =$

12. $\frac{1}{x} + \frac{2}{x^2} - \frac{3}{x^3} =$

Name________________________________ 7.EE.A.1

More Adding and Subtracting with Unlike Denominators

$$\frac{x+1}{x^2-9} + \frac{4}{x+3} - \frac{x-1}{x-3} = \frac{x+1}{(x+3)(x-3)} + \frac{4(x-3)}{(x+3)(x-3)} - \frac{(x-1)(x+3)}{(x-3)(x+3)}$$
$$= \frac{x+1+4x-12-(x^2+2x-3)}{(x+3)(x-3)}$$
$$= \frac{5x-11-x^2-2x+3}{(x+3)(x-3)}$$
$$= \frac{-x^2+3x-8}{x^2-9}$$

1. $\frac{3a+2b}{3b} - \frac{a+2b}{6a} =$

2. $\frac{a}{2a+2b} - \frac{b}{3a+3b} =$

3. $\frac{3x}{2y-3} + \frac{2x}{3-2y} =$

Hint: $3 - 2y = {}^{-}1(2y - 3)$

4. $\frac{x}{x+3} + \frac{9x+18}{x^2+3x} =$

5. $\frac{x+3}{x-5} + \frac{x-5}{x+3} =$

6. $\frac{11x}{x^2+3x-28} + \frac{x}{x+7} =$

7. $\frac{d^2+3}{d^2-2d} - \frac{d-4}{d} =$

8. $\frac{4a}{2a+6} - \frac{a-1}{a+3} =$

9. $\frac{a+b}{ax+ay} - \frac{a+b}{bx+by} =$

10. $\frac{8}{c^2-4} + \frac{2}{c^2-5c+6} =$

11. $\frac{x}{x^2-16} + \frac{6}{4-x} - \frac{1}{x-4} =$

12. $\frac{1}{a^2-a-2} + \frac{1}{a^2+2a+1} =$

13. $\frac{5}{3x-3} + \frac{x}{2x+2} - \frac{3x^2}{x^2-1} =$

14. $\frac{x+1}{x^2-9} + \frac{4}{x+3} - \frac{x-1}{x-3} =$

Name________________________________

8.EE.C.7b

Simplifying Mixed Expressions

$$\frac{a}{x+3} + \frac{a}{x-3} - 2 = \frac{a(x-3)}{(x+3)(x-3)} + \frac{a(x+3)}{(x+3)(x-3)} - \frac{2(x+3)(x-3)}{(x+3)(x-3)}$$

$$= \frac{ax - 3a + ax + 3a - 2x^2 + 18}{(x+3)(x-3)}$$

$$= \frac{2ax - 2x^2 + 18}{x^2 - 9}$$

1. $b + \dfrac{6}{b-1} =$

2. $3 + \dfrac{a+2b}{a-b} =$

3. $x - y + \dfrac{1}{x+y} =$

4. $7 + \dfrac{3}{a} + \dfrac{6}{b} =$

5. $\dfrac{5}{x+2} + 1 =$

6. $d + 3 + \dfrac{2d-1}{d-2} =$

7. $\dfrac{2x-3}{x+2} - 4 =$

8. $2x - \dfrac{x+y}{y} =$

9. $\dfrac{8}{3a-1} - 6 =$

10. $(x-4) - \dfrac{1}{x-2} =$

11. $\dfrac{x}{2y} - (x+2) =$

12. $(a+2) + \dfrac{7}{a-2} =$

13. $4 - \dfrac{3}{y-1} - \dfrac{1}{y+1} =$

14. $\dfrac{\frac{a}{b}+1}{\frac{a}{b}-1} =$

Name________________________________ 8.EE.C.7b

Synthetic Division

$$\frac{x^2 - 5x - 24}{x + 3} \Longrightarrow$$

$^-3$	1	$^-5$	$^-24$
		$^-3$	24
	1	$^-8$	0

$= x - 8$

Step 1. Use the opposite ($^-3$) of the constant term of the divisor ($x + 3$).
Step 2. Write the coefficients (1, $^-5$ $^-24$) of the dividend ($x^2 - 5x - 24$).
Step 3. Bring down, below the line, the first coefficient (1).
Step 4. Multiply the divisor ($^-3$) by the result of step 3. ($^-3 \cdot 1$)
Step 5. Place the product from step 4 ($^-3$) under the second coefficient ($^-5$) and add.
Step 6. Repeat steps 3 through 5, multiply, and add.
Step 7. Use the results to write the quotient.
(First term will have an exponent 1 less than the dividend.)

1. $y^2 - 13y + 36 \div y - 4 =$

Hint: 4 | 1 $^-13$ 36

2. $x^2 + 10x + 21 \div x + 3 =$

3. $4a^2 + 19a + 21 \div a + 1 =$

4. $x^3 - 5x^2 + 2x + 8 \div x - 2 =$

5. $y^2 + 25 \div y + 5 =$

6. $x^3 + 2x^2 - 2x + 24 \div x + 4 =$

Name________________________________

8.EE.C.7b

Solving Fractional Equations

$$\frac{2}{3x} + \frac{1}{2} = \frac{3}{4x}$$

$$12x \cdot \frac{2}{3x} + 12x \cdot \frac{1}{2} = 12x \cdot \frac{3}{4x}$$

$$8 + 6x = 9$$

$$6x = 1$$

$$x = \frac{1}{6}$$

1. $\frac{5}{6x} + 3 = \frac{1}{2x}$

2. $\frac{2}{5n} = \frac{3}{10n} - \frac{3}{5}$

3. $\frac{4}{3x} - \frac{5}{2x} = 5 + \frac{1}{6x}$

4. $\frac{c - 7}{c + 2} = \frac{1}{4}$

5. $\frac{y}{y - 3} = 2$

6. $\frac{2x}{5} + \frac{1}{2} = \frac{3x}{10}$

7. $\frac{x}{x - 2} = \frac{4}{5}$

8. $\frac{2}{3} = \frac{y}{y + 3}$

9. $\frac{10}{x - 3} = \frac{9}{x - 5}$

10. $\frac{7}{x} - \frac{4x}{2x - 3} = {}^{-}2$

11. $\frac{3}{x} + \frac{1}{2x} = \frac{7}{8}$

12. $\frac{3}{4} = \frac{x + 5}{x - 2}$

Name____________________________________

More Solving Fractional Equations

$$\frac{2}{x^2 - x} - \frac{2}{x - 1} = 1$$

$$x(x-1) \cdot \frac{2}{x(x-1)} - x(x-1)\frac{2}{x-1} = x(x-1) \cdot 1$$

$$2 - 2x = x^2 - x$$

$$0 = x^2 + x - 2$$

$$0 = (x + 2)(x - 1)$$

$$x + 2 = 0 \qquad x - 1 = 0$$

$$x = {}^{-}2 \qquad x = 1$$

1 is rejected because denominator ≠ 0.

1. $\frac{1}{u + 4} + \frac{1}{u - 4} = \frac{6}{u^2 - 16}$

2. $\frac{x}{8} + \frac{1}{x - 2} = \frac{x + 2}{2x - 4}$

3. $\frac{5y}{y + 1} - \frac{y}{y + 6} = 4$

4. $\frac{d}{d - 2} = \frac{d + 3}{d + 2} - \frac{d}{d^2 - 4}$

5. $\frac{6y}{2y + 1} - \frac{3}{y} = {}^{-}1$

6. $2 + \frac{4}{b - 1} = \frac{4}{b^2 - b}$

7. $\frac{2z^2 + z - 3}{z^2 + 1} = 2$

8. $\frac{x}{x - 3} + \frac{2}{x + 4} = 1$

9. $\frac{1}{m - 3} + \frac{1}{m + 5} = \frac{m + 1}{m - 3}$

10. $\frac{c}{c + 1} + \frac{3}{c - 3} + 1 = 0$

11. $\frac{b}{b + 1} - \frac{b + 1}{b - 4} = \frac{5}{b^2 - 3b - 4}$

12. $\frac{2}{2y + 1} - \frac{1}{2y} = \frac{3}{2y + 1}$

Name________________________________ 7.NS.A.3

Order of Operations

$2 + (2^2 + 6) \div {}^-2 - 1 = 2 + (4 + 6) \div {}^-2 - 1 = 2 + 10 \div {}^-2 - 1 = 2 + {}^-5 - 1 = {}^-4$

1. $(12 - 8) + 3 =$

2. $2 \cdot 6 + 4 \cdot 5 =$

3. $25 \div 5 \cdot 4 - 15 \cdot 8 =$

4. $3 + 15 \div 3 - 4 =$

5. $15 \div (7 - 2) + 3 =$

6. $2\,(7 + 3) \div 4 =$

7. $7 - (8 \cdot 2) \cdot 0 =$

8. $2\,[4 + (6 \div 2)] =$

9. $20 \div [2 + (7 - 4)] =$

10. $6\,({}^-9 + 4) \div 3 - 1 =$

11. $6 - 4\,(6 + 2) =$

12. $12 \div [(8 \div 2) \cdot (3 \div 3)] =$

13. $\dfrac{9^2 - 11}{(3 + 4) \cdot 10} =$

14. $\dfrac{3^2 - 4 \cdot 3 + 4}{9^2 - 4} =$

15. $\dfrac{3 \cdot 2 \div 6 + 2 \cdot 3 \div 6}{3^2 + 2^2 + 1^2} =$

16. $\dfrac{2 \cdot 4 - 6\,(2 + 1)}{1^2 - 3 \cdot 2} =$

17. $\dfrac{(4 - 6)^2}{{}^-24 \div 12} =$

18. $\dfrac{2 \cdot 6 - (4 + 2)}{({}^-2 - 4 - 6) \div (2 - 1)} =$

19. $\dfrac{{}^-3\,(4 - 9)}{35 \div {}^-7} =$

20. $3^5 \div 3^2 \div 3^2 \div 3 =$

Name________________________________

Solving Equations with Addition and Subtraction

$$16 + x = {}^{-}14$$
$$16 + {}^{-}16 + x = {}^{-}14 + {}^{-}16$$
$$x = {}^{-}30$$

1. $x + 7 = {}^{-}13$
2. $x + 7 = 4$
3. ${}^{-}14 + y = {}^{-}17$
4. $y - 11 = 14$
5. $y - 5 = {}^{-}7$
6. ${}^{-}20 + x = {}^{-}80$
7. $6 + x = 29$
8. $a + 32 = {}^{-}4$
9. ${}^{-}2 = x - 2$
10. ${}^{-}19 + y = 42$
11. $16 = z - 10$
12. $y + 73 = 0$
13. ${}^{-}100 = b + ({}^{-}72)$
14. $w - 5 = (8 - 13)$
15. $x + 2.5 = {}^{-}4.7$
16. $a + 3.6 = {}^{-}0.2$
17. $x - 6\frac{1}{4} = 12\frac{1}{2}$
18. $2\frac{1}{5} + x = {}^{-}3\frac{1}{2}$
19. $n + \frac{1}{2} = \frac{3}{4}$
20. $b - 1\frac{1}{3} = {}^{-}3\frac{5}{6}$

Name________________________________ 8.EE.C.7b, HSA-REI.B.3

Solving Equations with Multiplication and Division

$$2x = 12$$
$$\frac{2x}{2} = \frac{12}{2}$$
$$x = 6$$

$$-\frac{3}{4}y = 15$$
$$-\frac{4}{3} \cdot -\frac{3}{4}y = 15 \cdot -\frac{4}{3}$$
$$y = {}^{-}20$$

1. $3x = {}^{-}21$

2. ${}^{-}7y = 28$

3. ${}^{-}28 = {}^{-}196x$

4. ${}^{-}15a = {}^{-}45$

5. $-x = 17$

6. ${}^{-}21 = {}^{-}2x$

7. ${}^{-}12b = {}^{-}288$

8. $12x = {}^{-}60$

9. $\frac{a}{5} = {}^{-}6$

10. $-\frac{2}{5}y = {}^{-}14$

11. $\frac{3x}{4} = {}^{-}24$

12. $-\frac{x}{3} = \frac{4}{9}$

13. $-\frac{3}{7} = \frac{a}{14}$

14. $3a = -\frac{1}{4}$

15. $\frac{a}{2.4} = 0.26$

16. $-\frac{1}{99}y = 0$

17. ${}^{-}1.5x = 6$

18. ${}^{-}12.5 = 4n$

19. ${}^{-}3.7w = {}^{-}11.1$

20. $\frac{y}{6} = -\frac{2}{3}$

Name______________________________ 8.EE.C.7b, HSA-REI.B.3

Solving Basic Combined Equations

$$7\,(x + 2) = {}^{-}35$$
$$7x \quad 14 = {}^{-}35$$
$$7x + 14 - 14 = {}^{-}35 - 14$$
$$\frac{7x}{7} = \frac{{}^{-}49}{7}$$
$$x = {}^{-}7$$

1. $5x - 3 = 22$

2. $4a + 3 = {}^{-}5$

3. $5 - 7y = 33$

4. $5x - 11 = {}^{-}16$

5. ${}^{-}3 = 5x + 12$

6. $0 = 0.6x - 3.6$

7. $6 - 8x = {}^{-}26$

8. $5 = 5x + 27$

9. $3\,(w + 3) = {}^{-}15$

10. $2\,(y + 1) - 5 = 7$

11. $6 - \frac{2}{3}x = {}^{-}8$

12. $0.3x - 4.2 = 2.7$

13. $8.6 = 2.1 - 1.3y$

14. $5 - 4\,(y + 1) = {}^{-}3$

15. ${}^{-}1 = \frac{{}^{-}y}{4} - 6$

16. $\frac{5x}{6} + 34 = 9$

17. $\frac{{}^{-}2}{3}d + 3 = 11$

18. $1.2x + 6 = {}^{-}1.2$

19. $28 = \frac{17}{32}x - 23$

20. $\frac{2x}{5} + 4 = {}^{-}12$

Name________________________________ 8.EE.C.7b, HSA-REI.B.3

An Equation for Success

Solve each equation. Write the letter associated with the solution from the Answer Bank above the problem number below.

1. $y - 16 - 3y = 0$
2. $y - 4 - 4y = -1$
3. $8y + 4 - 2y = 22$
4. $12 = 3y - y + 4y$
5. $2(y - 3) - y = -1$
6. $3y - 2(y + 4) = 8$
7. $2(y - 1) - y = -2$
8. $5(2y - 2) + 4 = 4$
9. $3(y + 2) + 2(y - 3) = -15$
10. $4y - 2(y - 5) = -2$
11. $3y + 3(1 - y) - y = -6$
12. $7y - 3(y + 6) = 10$
13. $\frac{6y}{8} + 12 = 84$
14. $-2(-3 - 4y) = -10$
15. $-6y + 9 + 4y = -3$

ANSWER BANK

2	-2	4	1	-7	8	5	-5	-8	96
I	T	A	Y	Q	N	U	L	H	P
-8	-9	6	-3	-1	9	-4	-10	-6	10
O	F	T	!	R	M	B	S	R	W
15	-18	35	7	20	16	-16	3	-15	0
C	S	K	H	G	U	E	P	D	M

___ ___ ___ + ___ ___ ___ ___ ___ = ___ ___ ___ ___ ___ ___ ___

14 10 8   5 7 3 1 9   15 2 4 6 11 13 12

Name________________________________

8.EE.C.7b, HSA-REI.B.3

Solving Equations with Variables on Both Sides

$$4x - 6 = x + 9$$
$$4x - x - 6 = x - x + 9$$
$$3x - 6 = 9$$
$$3x - 6 + 6 = 9 + 6$$
$$\frac{3x}{3} = \frac{15}{3}$$
$$x = 5$$

1. $4x - 6 = x + 9$
2. $4 - 7x = 1 - 6x$
3. $^-4x - 3 = {}^-6x + 9$
4. $41 - 2n = 2 + n$
5. $6\ (2 + y) = 3\ (3 - y)$
6. $4y = 2\ (y - 5) - 2$
7. $6x - 9x - 4 = {}^-2x - 2$
8. $-(x + 7) = {}^-6x + 8$
9. $3 - 6a = 9 - 5a$
10. $^-9x + 6 = -x + 4$
11. $5x - 7 = {}^-10x + 8$
12. $y + 3 = 4y - 18$
13. $^-3\ (y + 3) = 2y + 3$
14. $2\ ({}^-3a + 5) = {}^-4\ (a + 4)$
15. $7x - 3 = 2\ (x + 6)$
16. $^-6x + 9 = 4\ (5 - x)$
17. $3\ (x + 2) = {}^-5 - 2\ (x - 3)$
18. $2\ (x - 3) = (x - 1) + 7$
19. $\frac{1}{3}\ (6y - 9) = {}^-2y + 13$
20. $\frac{1}{6}\ (12 - 6x) = 5\ (x + 4)$

Name________________________________ 8.EE.C.7b, HSA-REI.B.3

Solving Equations: Go with the Flow

Equation Flow Chart

1. Are there grouping symbols?
   No → Yes → Distribute.
2. Are there variables on the right side?
   No → Yes → Use inverse operation to move them to the left side.
   ↓ Simplify each side. ↓ Simplify each side.
3. Are there any constants (numbers without variables) on the left side?
   No → Yes → Use inverse operation to move them to the right side.
4. Is there a coefficient (number attached to the variable)?
   No → Yes → If by multiplication, then divide both sides.
   If by division, then multiply both sides.
   If a fraction, multiply by reciprocal on both sides.
5. Variable = number

Use the chart to solve each problem.

1. $3\,(x - 5) = 21$

2. $x + 30 = 4x - 6$

3. $^{-}6 + 2x = 9 - 3x$

4. $^{-}6x + 9 = {}^{-}4x - 3$

5. $\frac{3x}{5} = \frac{2x}{5} + 10$

6. $\frac{3x}{8} + 8 = {}^{-}40$

Name________________________________ 8.EE.C.7b, HSA-REI.B.3

Equation SUMmary

Solve the equations.

1. $\frac{3x}{2} - 9 = 0$

2. $6x + 3 = {}^{-}5x + 14$

3. $\frac{x}{8} + 3 = 2$

4. $5y = 2y - 42 - 3y$

5. $37 + 8x = 4\ (7 - x)$

6. $5\ (2 - x) = 7x - 26$

7. $6 + 4x = \frac{(6x + 9)}{3}$

8. $1.6\ (3y - 1) + 2 = 5y$

9. $7x - 10 = 6\ (11 - 2x)$

10. $3\ (4x - 9) = 5\ (2x - 5)$

11. $(\frac{2}{3})\ (x + 9) = x + 5$

12. $\frac{5y}{7} - 15 = 5y + 30$

CHECK: sum of the x solutions $= -\frac{1}{2}$ sum of the y solutions.

____________ $= -\frac{1}{2}$ ____________

Name________________________________ (8.EE.C.7b, HSA-CED.A.1, HSA-REI.B.3)

Problem Solving Using Equations

Set up and solve each equation.

The sum of twice a number and 21 is 83. Find the number.	$2n + 21 = 83$ $2n + 21 - 21 = 83 - 21$ $2n = 62$ $n = 31$ The number is 31.

1. Twice a number, diminished by 17 is $^-3$. Find the number.

2. Six times a number, increased by 3 is 27. Find the number.

3. Three times the difference of 5 minus a number is 27. Find the number.

4. Karl's team score is 39 points. This was one point less than twice Todd's team score. Find Todd's team score.

5. The length of a rectangle is 6 feet more than twice the width. If the length is 24 feet, what is the width?

6. Four fifths of the third grade went on a trip to the zoo. If 64 children made the trip, how many children are in the third grade?

7. The price of a pack of gum today was 63¢. This is 3¢ more than three times the price ten years ago. What was the price ten years ago?

8. The sum of three consecutive integers is 279. Find the integers.

9. The sum of two consecutive odd integers is 112. Find the integers.

10. Find four consecutive integers such that the sum of the second and fourth is 132.

11. Find three consecutive odd integers such that their sum decreased by the second equals 50.

Name________________________________ 8.EE.C.7b, HSA-CED.A.1, HSA-REI.B.3

More Problem Solving Using Equations

Set up and solve each equation.

The sum of two numbers is 52. The difference of the same two numbers is 20. Find the numbers.

x = one number

$$x - (52 - x) = 20$$
$$x - 52 + x = 20$$
$$2x - 52 = 20$$
$$2x - 52 + 52 = 20 + 52$$
$$\frac{2x}{2} = \frac{72}{2}$$
$$x = 36$$

$52 - x$ = second number

$52 - x = 52 - 36 = 16$

The numbers are 36 and 16.

1. One number is four times another. Their sum is 35. Find the numbers.

2. The sum of two numbers is 21. One number is three less than the other. Find the numbers.

3. The greater of two numbers is one less than 8 times the smaller. Their sum is 98. Find the numbers.

4. In a triangle, the second angle measures twice the first, and the third angle measures 5 more than the second. If the sum of the angles' measures is 180°, find the measure of each angle.

5. The length of a rectangle is 4 centimeters (cm) less than three times the width. The perimeter is 64 cm. Find the width and length. (**Hint:** Perimeter = 2l + 2w)

6. The sum of three numbers is 64. The second number is 3 more than the first. The third number is 11 less than twice the first. Find the numbers.

7. Bill can type 19 words per minute faster than Bob. Their combined typing speed is 97 words per minute. Find Bob's typing speed.

Name________________________________ (8.EE.C.7b, HSA-REI.B.3)

Equations: A Little Magic

Set up and solve each equation. In a magic square, each row, column, and diagonal has the same sum.

$3x - 5 = 2(2x + 5)$	$10x + 8 = 12x - 18$	$4(x - 7) = 2x - 6$	$3(y + 4) = 5y + 30$
$3(2y + 4) = 4(y + 7) - 2$	$^{-}6y = 10 - 4y$	$^{-}27 - 6x = 3x$	$7y + 3 = 12y - 2$
$2y + 2 = 3y + 3$	$8x - (6x - 4) = 10$	$6y = 10 + 4y$	$6(x + 7) = 2(x + 7)$
Two consecutive whole numbers total 17. Find the larger.	Two consecutive odd numbers total $^{-}20$. Find the smaller.	One number is 4 less than 3 times another. Their sum is $^{-}16$. Find the smaller.	Two consecutive odd numbers total 32. Find the smaller.

Magic Sum is ______________ .

Name________________________

8.EE.C.7b, HSA-REI.B.3

Solving Equations with Negative Variables

$$\frac{^-k}{6} + 1 = {^-5}$$

$$\frac{^-k}{6} + 1 - 1 = {^-5} - 1$$

$$\frac{^-k}{6} \cdot 6 = {^-6} \cdot 6$$

$${^-k} = {^-36}$$

$$k = 36$$

1. ${^-8} - y = 22$

2. $18 = {^-k} + 3$

3. $4 - \frac{x}{5} = {^-16}$

4. $-x - 15 = {^-15}$

5. $-z = 11$

6. ${^-28} = \frac{^-y}{4} - 12$

7. ${^-82} = -a$

8. $\frac{^-b}{3} + 50 = 100$

9. ${^-6} - x\frac{1}{9} = {^-18}$

10. ${^-3z} + 5 = 38$

11. $-a\frac{1}{2} + 12 = {^-9}$

12. ${^-5y} - {^-7} = 52$

Name________________________________

8.EE.C.7b, HSA-REI.B.3

Solving Equations—Variables on Both Sides

$$5x + 6 = 2x + 5$$
$$5x - 2x + 6 = 2x - 2x + 15$$
$$3x + 6 - 6 = 15 - 6$$
$$\frac{3x}{3} = \frac{9}{3}$$
$$x = 3$$

1. $20y + 5 = 5y + 65$

2. $13 - t = t - 7$

3. $^{-}3k + 10 = k + 2$

4. $^{-}9r = 20 + r$

5. $6m - 2\frac{1}{2} = m + 12\frac{1}{2}$

6. $18 + 4.5p = 6p + 12$

7. $5x - \frac{1}{4} = 3x - \frac{5}{4}$

8. $^{-}x - 2 = 1 - 2x$

9. $3k + 10 = 2k - 21$

10. $8y - 6 = 5y + 12$

11. $^{-}t + 10 = t + 4$

12. $4m - 9 = 5m + 7$

Name_______________________

Solving Problems with Two Variables

Set up and solve each equation.

If 8 pens and 7 pencils cost \$3.37 while 5 pens and 11 pencils cost \$3.10, how much does each pen and pencil cost?

Let x = cost of 1 pen. Let y = cost of 1 pencil.

$8x + 7y = 337$

$5x + 11y = 310$

$^{-}40x - 35y = {}^{-}1{,}685$

$40x + 88y = 2{,}480$

$53y = 795$

$y = 15$

$8x + 7 \cdot 15 = 337$

$8x + 105 = 337$

$8x = 232$

$x = 29$

Pens cost \$.29 and pencils cost \$.15.

1. A rectangle has a perimeter of 18 cm. Its length is 5 cm greater than its width. Find the dimensions.

2. Timmy has 180 marbles, some plain and some colored. If there are 32 more plain marbles than colored marbles, how many colored marbles does he have?

3. A theater sold 900 tickets to a play. Floor seats cost \$12 each and balcony seats \$10 each. Total receipts were \$9,780. How many of each type of ticket were sold?

4. Ryan and Karl spent 28 hours building a tree house. Ryan worked 4 more hours than Karl. How many hours did each work?

5. The difference between seven times one number and three times a second number is 25. The sum of twice the first and five times the second is 95. Find the numbers.

6. The sum of two numbers is 36. Their difference is 6. Find the numbers.

7. The volleyball club has 41 members. There are 3 more boys than girls. How many girls are there?

8. The sum of two numbers is 15. Twice one number equals 3 times the other. Find the numbers.

Name________________________________ 8.EE.C.8a, HSA-CED.A.2

Solving for *y*

Solve each equation for *y*. Then use the given values for *x* to find the corresponding values for *y*. Write answers as ordered pairs.

$$y - 4 = 3x$$
$$y - 4 + 4 = 3x + 4$$
$$y = 3x + 4$$

Let $x = {}^-2, 0, 1$

Solve for *y*

a. $y = 3 \cdot {}^-2 + 4$
$y = {}^-6 + 4$
$y = {}^-2$
$({}^-2, {}^-2)$

b. $y = 3 \cdot 0 + 4$
$y = 0 + 4$
$y = 4$
$(0, 4)$

c. $y = 3 \cdot 1 + 4$
$y = 3 + 4$
$y = 7$
$(1, 7)$

1. $y = 5x$ Let $x = {}^-3, 0, 2$

Note: This equation is already in the form of $y = ...$

2. $2x + y = 9$ Let $x = {}^-1, 0, 5$

3. $-x = y + 3$ Let $x = {}^-3, 0, 4$

4. $y = \frac{2}{3}x + 1$ Let $x = {}^-4, 0, 3$

5. $8x + y = 1$ Let $x = {}^-2, 0, 1$

6. $y - 1 = {}^-3x$ Let $x = {}^-3, 0, 2$

7. $2 = y - \frac{1}{3}x$ Let $x = {}^-9, 0, 6$

8. $7x - y = {}^-8$ Let $x = {}^-1, 0, {}^-3$

Name________________________________

7.EE.A.1

Introduction to Factoring

Find the GCF of the numbers.

18, 30
$18 = 2 \cdot 3 \cdot 3$
$30 = 2 \cdot 3 \cdot 5$
$2 \cdot 3 = 6$
$6 = \text{GCF}$

1. 12, 18
2. 10, 35
3. 8, 30
4. 16, 24
5. 28, 49
6. 27, 63
7. 30, 45
8. 48, 72

Greatest Common Monomial Factor

Factor. Write prime if prime.

$12a^3b + 15ab^3 = 3ab\,(4a^2 + 5b^2)$

1. $6x + 3 =$
2. $24x^2 - 8x =$
3. $6x - 12 =$
4. $2x^2 + 8x =$
5. $4x + 10 =$
6. $10x^2 + 35x =$
7. $10x^2y - 15xy^2 =$
8. $12x^2 - 9x + 15 =$
9. $3n^3 - 12n^2 - 30n =$
10. $9m^2 - 4n + 12 =$
11. $2x^3 - 3x^2 + 5x =$
12. $13m + 26m^2 - 39m^3 =$
13. $17x^2 + 34x + 51 =$
14. $18m^2n^4 - 12m^2n^3 + 24m^2n^2 =$

Name________________________________ 7.EE.A.1, HSA-SSE.A.2

Factoring the Difference of Two Squares

$a^2 - 36 = (a + 6)(a - 6)$
$3x^2 - 48 = 3(x^2 - 16) = 3(x - 4)$

Factor. Write prime if prime.

1. $x^2 - 1 =$
2. $x^2 - 9 =$
3. $x^2 + 4 =$
4. $x^2 - 25 =$
5. $9y^2 - 16 =$
6. $4x^2 - 25 =$
7. $9x^2 - 1 =$
8. $a^2 - x^2 =$
9. $25 - m^2 =$
10. $x^2 - 16y^2 =$
11. $25m^2 - n^2 =$
12. $-x^2 + 16 =$
13. $36m^2 - 121 =$
14. $2x^2 - 8 =$
15. $25 + 4x^2 =$
16. $4a^2 - 81b^2 =$
17. $12x^2 - 75 =$
18. $a^2b - b^3 =$
19. $^-98 + 2x^2 =$
20. $5x^2 - 45y^2 =$
21. $9x^4 - 4 =$
22. $16x^4 - y^2 =$

Name________________________________ 7.EE.A.1, HSA-SSE.A.2

Factoring Challenge

Factor. Write prime if prime.

1. $a^2 - 36 =$

2. $9x^2 - 49 =$

3. $169m^2 - 4u^2 =$

4. $x^2y^2 - 9z^4 =$

5. $\frac{1}{4}x^2 - 25y^2 =$

6. $\frac{1}{9}x^2 - 16 =$

7. $64 - a^4b^4 =$

8. $y^6 - 100 =$

9. $\frac{4}{9}x^2y^2 - \frac{25}{36}z^2 =$

10. $y^8 - 81 =$

11. $1 - 8u + 16u^2 =$

12. $a^2b^2 + 6ab + 9 =$

13. $x^2 + 2xy + y^2 =$

14. $4x^2 + 12xy + 9y^2 =$

15. $100h^2 + 20h + 1 =$

16. $9a^2 - 24a + 16 =$

17. $4a^3 + 8a^2 + 4a =$

18. $5c + 20c^2 + 20c^3 =$

19. $(x + 4)^2 - (y + 1)^2 =$

20. $(x - 1)^2 - 10(x - 1) + 25 =$

Name________________________________

7.EE.A.1, HSA-SSE.A.2

Factoring Expressions

Factor each expression. Match the expression with its factored form. Write the answer letters above the problem numbers. Rewrite the familiar proverb.

1. $3x^2 + 4x + 1 =$
2. $5x^2 + 7x + 2 =$
3. $2x^2 - 11x + 5 =$
4. $3x^2 + x - 2 =$
5. $5x^2 - 2x - 7 =$
6. $8x^2 - 10xy + 3y^2 =$
7. $6x^2 + 19x + 15 =$
8. $28x^2 + 5xy - 12y^2 =$
9. $2x^2 + 7xy - 15y^2 =$
10. $12x^2 + 17x + 6 =$
11. $4x^2 - 4xy - 5y^2 =$
12. $56x^2 + 15y - 56 =$
13. $12x^2 - 29xy + 14y^2 =$
14. $64x^2 - 32xy - 21y^2 =$
15. $16x^2 + 56xy + 49y^2 =$
16. $18x^2 - 57x + 35 =$

A. $(7x - 4y)(4x + 3y)$
B. $(8x - 7y)(8x + 3y)$
C. $(7x + 8)(8x - 7)$
D. $(3x + 2)(4x + 3)$
E. $(2x - 3y)(x + 5y)$
H. Prime
I. $(2x + 3)(3x + 5)$
J. $(6x - 5)(3x - 7)$
L. $(3x - 2)(x + 1)$
N. $(2x - 1)(x - 5)$
O. $(5x - 7)(x + 1)$
P. $(4x - 7y)(3x - 2y)$
R. $(5x + 2)(x + 1)$
S. $(4x - 3y)(2x - y)$
T. $(3x + 1)(x + 1)$
U. $(4x + 7y)(4x + 7y)$

__ __ __ __ __ __ __ __ __　　__ __　　__ __ __ __ __ __ __ __ __ __ __
2 9 6 7 10 9 3 1 6　　7 3　　1 2 8 3 6 13 8 2 9 3 1

__ __ __ __ __ __　　__ __ __ __ __ __　　__ __ __　　__ __ __ __
8 14 5 10 9 6　　6 11 5 15 4 10　　3 5 1　　6 9 3 10

__ __ __ __ __ __ __ __ __ __ __
13 2 5 16 9 12 1 7 4 9 6

Proverb: ____.

Name________________________________

7.EE.A.1, HSA-SSE.A.2

Factoring: Putting It All Together

$5x^2 + 20x - 60 = 5\ (x^2 + 4x - 12) = 5\ (x + 6)\ (x - 2)$

Factor completely. Write prime if prime.

1. $2x^2 - 8 =$
2. $2x^2 + 8x + 6 =$
3. $3n^2 + 9n - 30 =$
4. $6x^2 - 26x - 20 =$
5. $2x^2 + 12x - 80 =$
6. $5t^2 + 15t + 10 =$
7. $8n^2 - 18 =$
8. $14x^2 + 7x - 21 =$
9. $4x^2 + 16x + 16 =$
10. $18x + 12x^2 + 2x^3 =$
11. $2x - 2xy^2 =$
12. $3t^3 - 27t =$
13. $24a^2 - 30a + 9 =$
14. $10x^2 + 15x - 10 =$
15. $3x^2 - 42x + 147 =$
16. $4x^4 - 4x^2 =$

Name________________________________ 7.EE.A.1, HSA-SSE.A.2

Factoring Summary

1. $16x^2 - 40x - 24 =$

2. $27x^2 - 36x + 12 =$

3. $5x^2 - 60x - 140 =$

4. $6m^3 + 54m^2 - 6m =$

5. $5k^4 + 8k^3 - 4k^2 =$

6. $x^2y^4 - x^6 =$

7. $y^4 - 6y^2 - 16 =$

8. $x^4 - 3x^2 - 4 =$

9. $h^2 - (a^2 - 6a + 9) =$

10. $81x^4 - 16y^4 =$

11. $4mn^2 - 4m^2n^2 + m^3n^2 =$

12. $(2a + 3)^2 - (a - 1)^2 =$

13. $16d^8 - 8d^4 + 1 =$

14. $x^2 (x^2 - 4) + 4x (x^2 - 4) + 4 (x^2 - 4) =$

Name______________________________

HSA-SSE.A.2, HSA-REI.B.3

Solving Equations Using Factoring

1. Rewrite equation in standard form (one member equals 0).
2. Factor completely.
3. Set each factor equal to 0; then solve.
4. Check results in original equation.

$x^2 - 7x + 12 = 0$
$(x - 4)(x - 3) = 0$
$x - 4 = 0$ or $x - 3 = 0$
$x = 4 \qquad x = 3$
$x = 3, 4$

$v^3 = 10v - 3v^2$
$v^3 + 3v^2 - 10v = 0$
$v(v^2 + 3v - 10) = 0$
$v(v + 5)(v - 2) = 0$
$v = 0$ or $v + 5 = 0$ or $v - 2 = 0$
$v = {}^-5 \qquad v = 2$
$v = {}^-5, 0, 2$

1. $x^2 - 5x - 6 = 0$
2. $v^3 - 4v = 0$
3. $n^2 - 16n = 0$
4. $x^2 + 9 = 10x$
5. $6x^2 = 16x - 8$
6. $s^2 = 56s - s^3$
7. $3y^2 + 2y - 1 = 0$
8. $u^3 = 14u^2 + 32u$
9. $23p = 5p^2 + 24$
10. $x^2 - 3x - 10 = 0$
11. $y^2 = 49$
12. $y^2 = {}^-7y - 10$
13. $x^2 = 8x$
14. $3x^2 - 2 = x^2 + 6$
15. $4y^2 = {}^-4y - 1$
16. $5x^2 - 2x - 3 = 0$

Name____________________________________ HSA-SSE.A.2, HSA-CED.A.1, HSA-REI.B.3

Factoring: Number Problems

Set up and solve each equation.

The sum of the squares of two consecutive, positive, even integers is 340. Find the integers.

Let x = 1st integer

$x + 2$ = 2nd integer

$$(x)^2 + (x + 2)^2 = 340$$
$$x^2 + x^2 + 4x + 4 = 340$$
$$2x^2 + 4x - 336 = 0$$
$$2(x^2 + 2x - 168) = 0$$
$$2(x + 14)(x - 12) = 0$$

$x + 14 = 0$ or $x - 12 = 0$

$x = -14$ rejected; $x = 12$

The integers are 12 and 14.

1. Fourteen less than the square of a number is the same as five times the number. Find the number.

2. When a number is added to six times its square, the result is 12. Find the number.

3. Find two consecutive, negative integers whose product is 156.

4. The sum of the squares of two consecutive integers is 41. Find the integers.

5. The sum of the squares of three consecutive, positive integers is equal to the sum of the squares of the next two integers. Find the five integers.

6. Find two consecutive even integers whose product is 80.

7. Twice the square of a certain positive number is 144 more than twice the number. What is the number?

8. The square of a positive number decreased by 10 is 2 more than 4 times the number. What is the number?

Name____________________

HSA-SSE.A.2, HSA-CED.A.1, HSA-REI.B.3

Factoring: Geometry Problems

Set up and solve each equation.

A square field had 9cm added to its length and 3cm added to its width. Its new area is $280cm^2$. Find the length of a side of the original field.

Let x = length of a side of the square field

$$(x + 9)(x + 3) = 280$$
$$x^2 + 12x + 27 = 280$$
$$x^2 + 12x - 253 = 0$$
$$(x + 23)(x - 11) = 0$$
$$x + 23 = 0 \text{ or } x - 11 = 0$$
$$x = {}^-23 \qquad x = 11$$

rejected

The side of the square was 11 cm.

1. The length of a rectangle is 5m greater than twice its width, and its area is $33m^2$. Find the dimensions.

2. The perimeter of a rectangular piece of property is 8 miles, and its area is 3 square miles. Find the dimensions. (**Hint**: $\frac{1}{2}$ P = l + w)

3. When the dimensions of a 2cm x 5cm rectangle were increased by equal amounts, the area was increased by $18cm^2$. Find the dimensions of the new rectangle.

4. If the sides of a square are increased by 3 in., the area becomes 64 in.^{2} Find the length of a side of the original square.

5. A rug placed in a 10 ft. x 12 ft. room covers two-thirds of the floor area and leaves a uniform strip of bare floor around the edges. Find the dimensions of the rug.

6. The area of a rectangular pool is 192 square meters. The length of the pool is 4 meters more than its width. Find the length and the width.

Name________________________________

Extra: Factoring by Grouping

$$6ax - 2b - 3a + 4bx = 6ax - 3a + 4bx - 2b$$
$$= 3a\ (2x - 1) + 2b\ (2x - 1)$$
$$= (2x - 1)\ (3a + 2b)$$

1. $x^2 + 2x + xy + 2y =$

2. $3a^2 - 2b - 6a + ab =$

3. $t^3 - t^2 + t - 1 =$

   Hint: $t - 1 = 1\ (t - 1)$

4. $10 + 2t - 5s - st =$

5. $\frac{2}{3}bc - \frac{14}{3}b + c - 7 =$

6. $4u^2 + v + 2uv + 2u =$

7. $ad + 3a - d^2 - 3d =$

8. $n^2 + 2n + 3mn + 6m =$

9. $2ax^2 + bx^2 - 2ay^2 - by^2 =$

10. $yz^2 - y^3 + z^3 - y^2z =$

11. $y^3 - y^2 - 4y + 4 =$

12. $x^2a + x^2b - 16a - 16b =$

13. $x^3 + x^2 - x - 1 =$

14. $a^3 - a^2 - 8a + 8 =$

Name_______________________________ HSA-SSE.A.2, HSA-CED.A.1, HSA-REI.B.3

From Factors to Equations

Since factored forms of equations can be used to determine the solutions, solutions can be used to determine the equation. Just reverse the solution process.

Example: If $^{-}1$ and $\frac{2}{3}$ are the solutions, what was the equation?

1. Use the solutions to write simple equations equal to zero. $x + 1 = 0$ and $3x - 2 = 0$
2. Use the expressions as factors of an equation. $(x + 1)(3x - 2) = 0$
3. Expand the factored form. $3x^2 - 2x + 3x - 2 = 0$
4. Simplify. $3x^2 + x - 2 = 0$

Use the given solutions to write equations. Shade the regions below containing the equations.

1. $x = 6, 4$
   1. ______
   2. ______
   3. ______
   4. ______
2. $x = {}^{-}7, {}^{-}2$
   1. ______
   2. ______
   3. ______
   4. ______
3. $x = 1, 5$
   1. ______
   2. ______
   3. ______
   4. ______
4. $x = \frac{1}{2}, 3$
   1. ______
   2. ______
   3. ______
   4. ______
5. $x = -\frac{3}{4}, 2$
   1. ______
   2. ______
   3. ______
   4. ______
6. $x = -\frac{1}{4}, {}^{-}8$
   1. ______
   2. ______
   3. ______
   4. ______
7. $x = 1, {}^{-}1$
   1. ______
   2. ______
   3. ______
   4. ______
8. $x = 0, 9$
   1. ______
   2. ______
   3. ______
   4. ______
9. $x = 0, {}^{-}2, {}^{-}3$
   1. ______
   2. ______
   3. ______
   4. ______

Three solutions mean three factors.

$4x^2 - 5x - 6 = 0$ P	$x^2 - 10x + 24 = 0$ A	$x^2 - 6x + 5 = 0$ R	$x^3 + 5x^2 + 6x = 0$ T	$x^2 - 9x - 14 = 0$ C
$x^2 - 1 = 0$ L	$4x^2 + 33x + 8 = 0$ M	$4x^2 + 5x + 6 = 0$ U	$x^2 - 9x = 0$ N	$2x^2 + x + 3 = 0$ B
$x^2 + x - 1 = 0$ I	$x^2 + 9x + 14 = 0$ E	$x^2 - 2x - 24 = 0$ A	$x^2 + 1 = 0$ C	$2x^2 - 7x + 3 = 0$ R

The remaining letters spell the name of the type of equation in problem 9: ______________

Name____

Just for Fun

Make your own matrix.

Maureen, Joan, Robert, and Bryan each have two favorite collections. The collections are seashells, stamps, baseball cards, coins, comic books, dolls, bugs, and rocks. No two children collect the same things. Find out the two collections each child has.

1. Maureen always finds things for both her collections outdoors.
2. Joan's friend enjoys collecting stamps.
3. One of Bryan's friends enjoys collecting coins.
4. The person who collects comics does not collect baseball cards.
5. One of Bryan's hobbies involves lots of reading.
6. Joan's family has a beach house; this is very helpful for one of her collections.
7. One of the girls collects dolls.

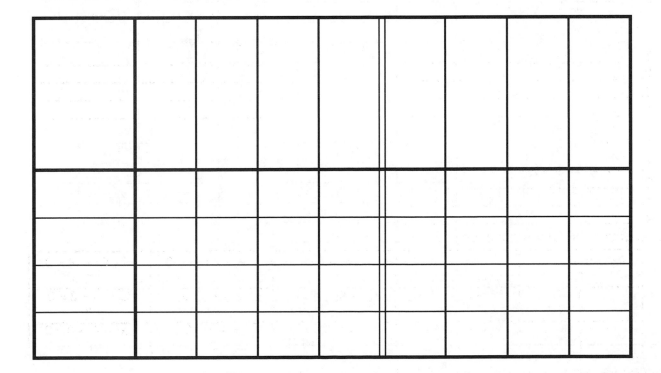

Name________________________________

HS8.F.A.1, HSA-REI.D.10, HSF-LE.A.2

Graphing Linear Equations by Plotting Points

Solve each equation for *y*. Then, choose 3 values for *x* and find the corresponding values for *y*. Graph the 3 ordered pairs and draw the line that contains them.

$$5x + y = {}^{-}1$$
$$5x - 5x + y = {}^{-}1 - 5x$$
$$y = -5x - 1$$

x	*y*
$^{-}1$	4
0	$^{-}1$
2	$^{-}11$

a. $y = {}^{-}5 \cdot {}^{-}1 - 1$
$y = 5 - 1$
$y = 4$

b. $y = {}^{-}5 \cdot 0 - 1$
$y = 0 - 1$
$y = {}^{-}1$

c. $y = {}^{-}5 \cdot 2 - 1$
$y = {}^{-}10 - 1$
$y = {}^{-}11$

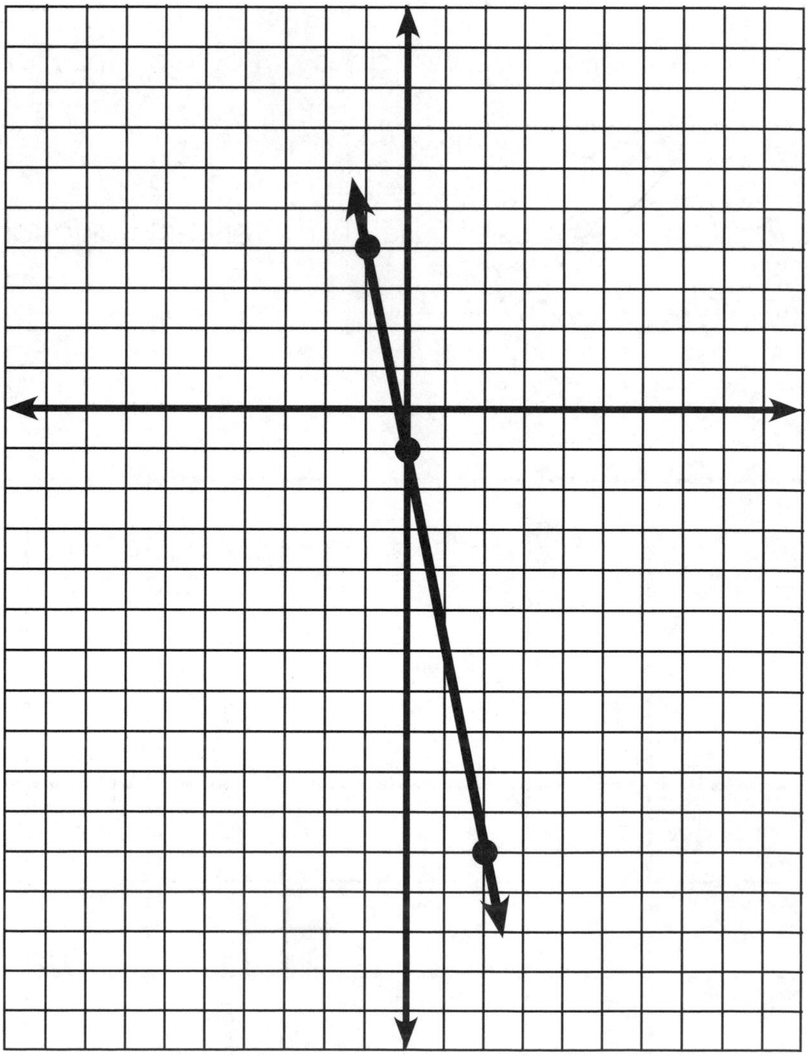

Name________________________________

HS8.F.A.1, HSA-REI.D.10

Graphing Linear Equations by Plotting Points

1. $y = \frac{1}{2}x - 3$

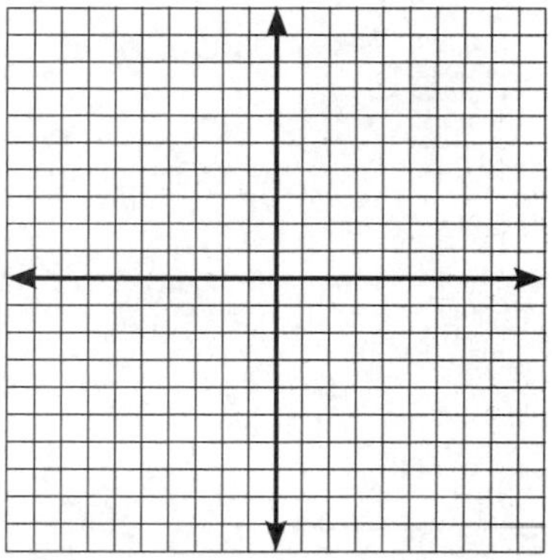

2. $^{-}2x + y = 5$

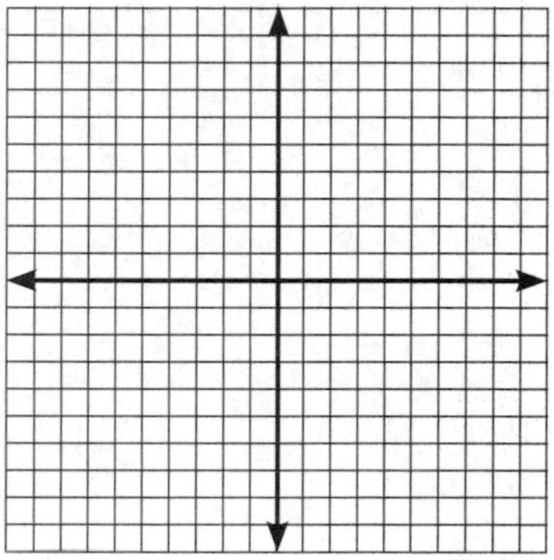

3. $4x + y = {}^{-}7$

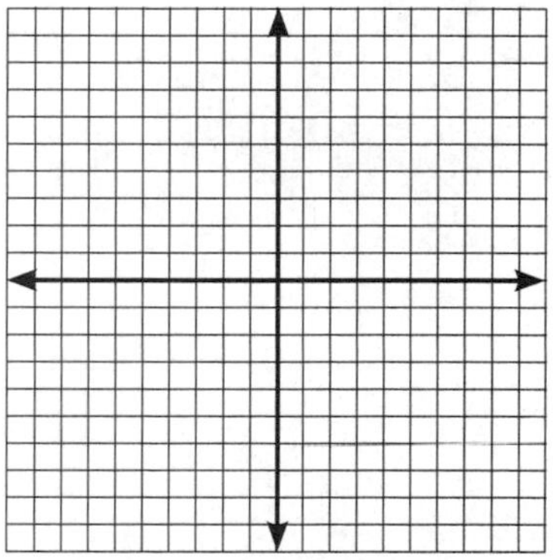

4. $y - 3 = 2x$

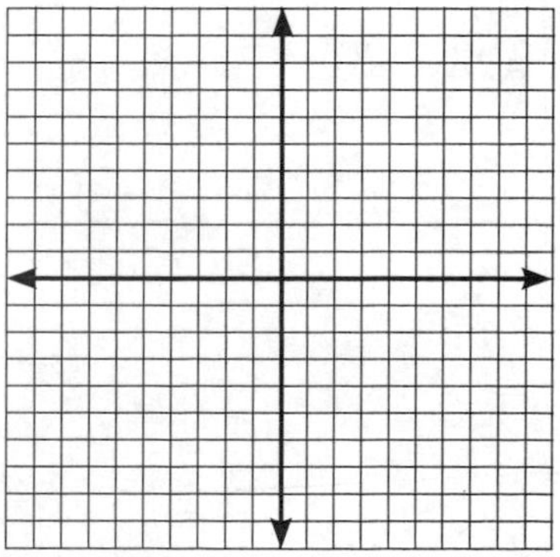

5. $3x + y = 7$

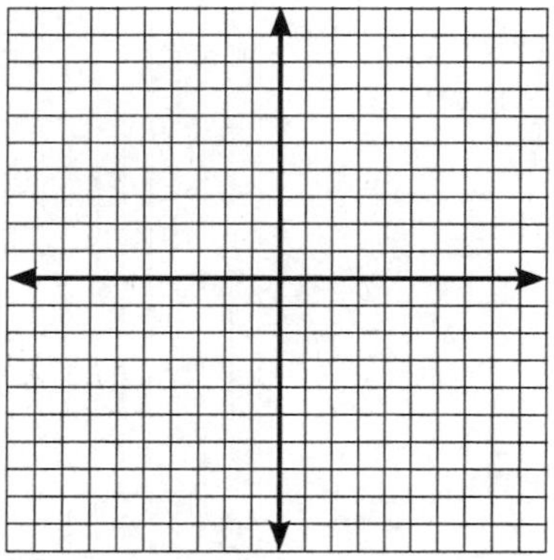

6. $3x - y = {}^{-}2$

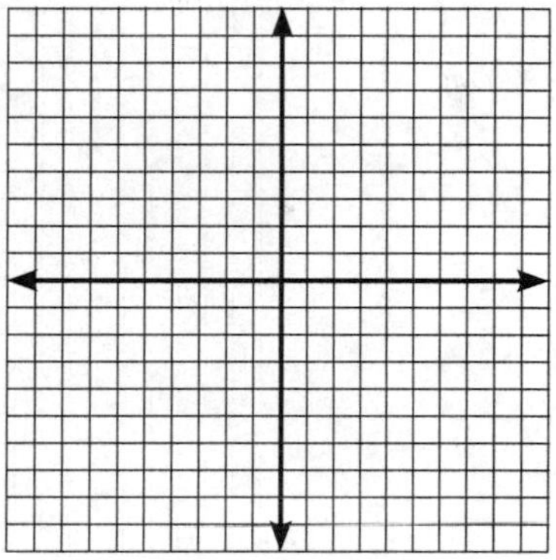

7. $y + 7 = 5x$

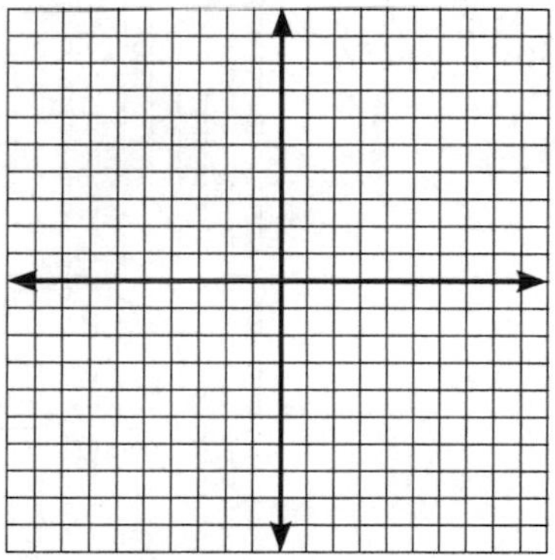

8. $y = \frac{1}{4}x - 2$

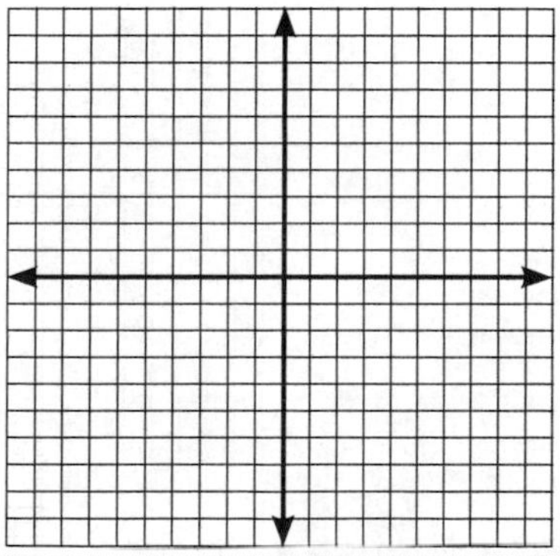

Name________________________________

8.F.A.1, HSA-REI.D.10

Graphing Linear Equations by Plotting Points

9. $\frac{3}{4}x + y = 2$

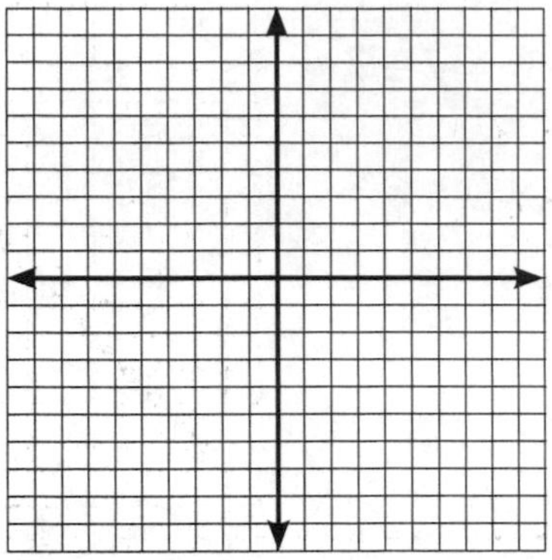

10. $x - y = {}^{-}4$

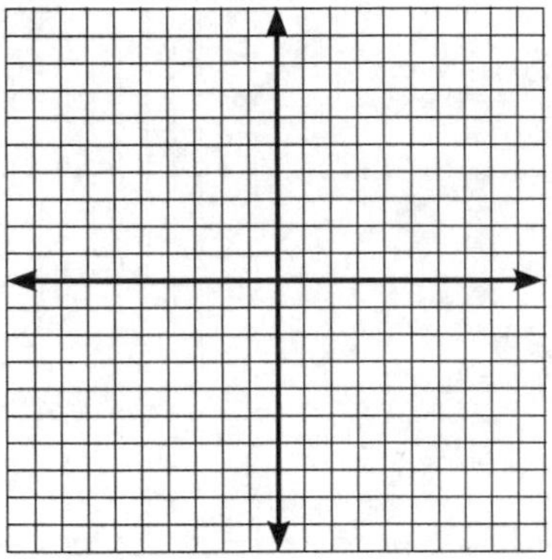

11. ${}^{-}3x + y = {}^{-}4$

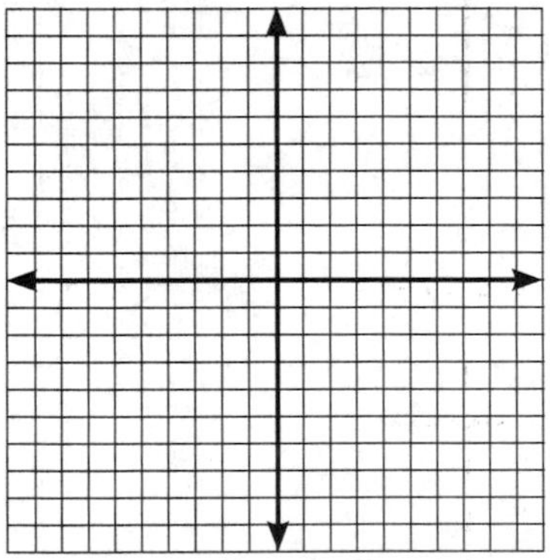

12. $y - x = {}^{-}1$

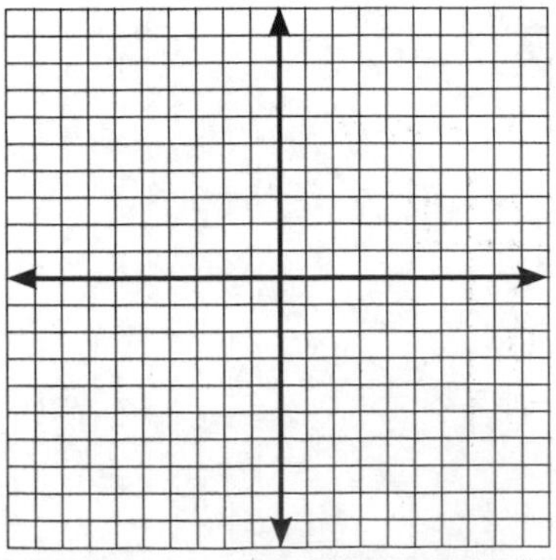

13. $y = -\frac{1}{2}x$

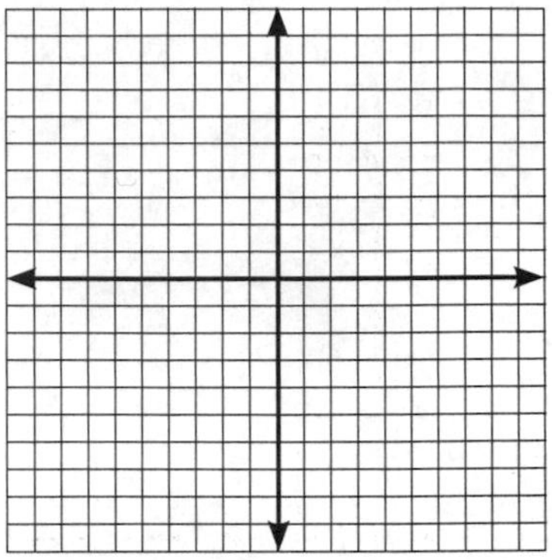

14. $\frac{1}{4}x + y = {}^{-}2$

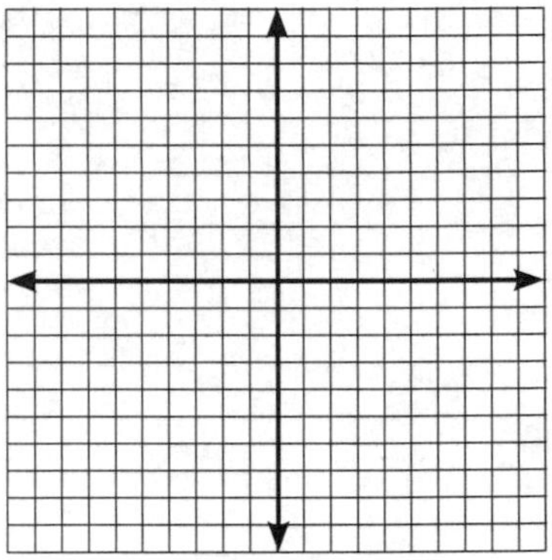

15. $y = {}^{-}4x - 5$

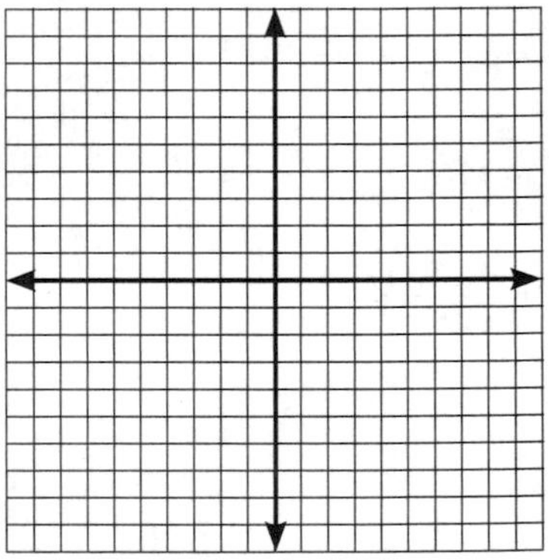

16. $y = {}^{-}2x - 3$

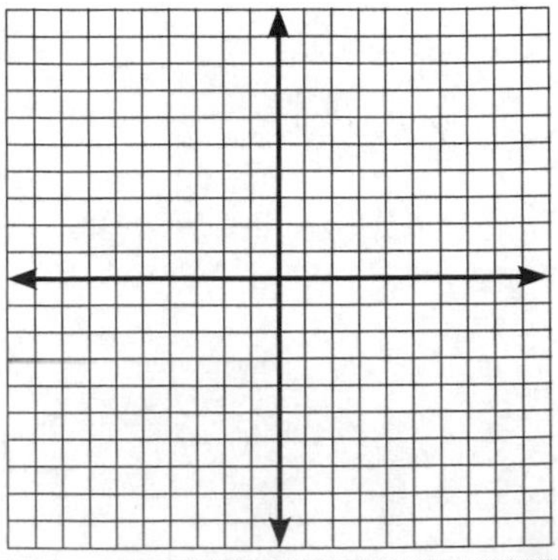

Name________________________________

8.F.A.1, HSA-REI.D.10, HSA-REI.D.11

Graphing Equations

Graph each equation by plotting points. Use your own graph paper.

$y = x + 2$

x	y
3	5
0	2
$^-2$	0

$y = 3 + 2$ $\quad$ $y = 0 + 2$ $\quad$ $y = {^-2} + 2$

$y = 5$ $\quad$ $y = 2$ $\quad$ $y = 0$

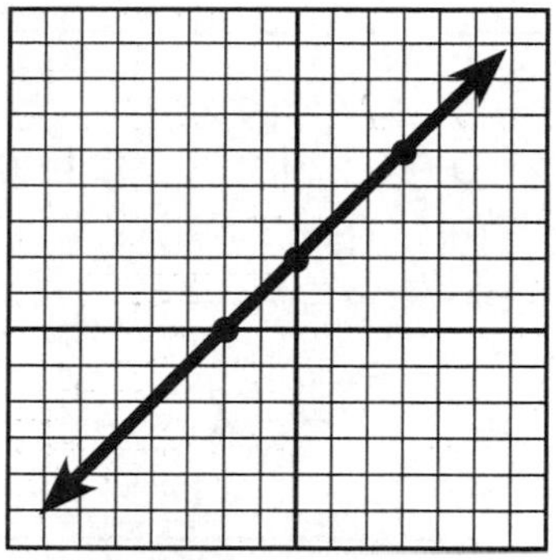

1. $y = x + 3$

2. $y = {^-2}x + 4$

3. $x + y = 2$

4. $3x + y = 9$

5. $x = {^-7}$

6. $y = {^-3}x + 5$

x	y
2	
0	
$^-1$	

7. $y = 5 - x$

x	y
7	
3	
0	

8. $3x + 4y = 12$

x	y
4	
0	
$^-4$	

Name____ 8.F.A.1, HSA-REI.D.10, HSA-REI.D.11

x and *y* Intercepts

I. Find the *x* and *y* intercepts.

$$2x + y = 3$$

To find *x*-intercept, let $y = 0$. To find y-intercept, let $x = 0$.

$2x + 0 = 3$

$2x = 3$

$x = \frac{3}{2}\ (\frac{3}{2}, 0)$

$2 \cdot 0 + y = 3$

$y = 3\ (0, 3)$

1. $3x + 4y = 12$
2. $4x + y = 2$
3. $5x - 4y = 15$
4. $2x - 2y = {}^-4$
5. $3x + y = {}^-9$
6. $4x - 2y - 8 = 0$

II. Find the *x* and *y* intercepts. Then, graph.

7. $x + 2y = 5$

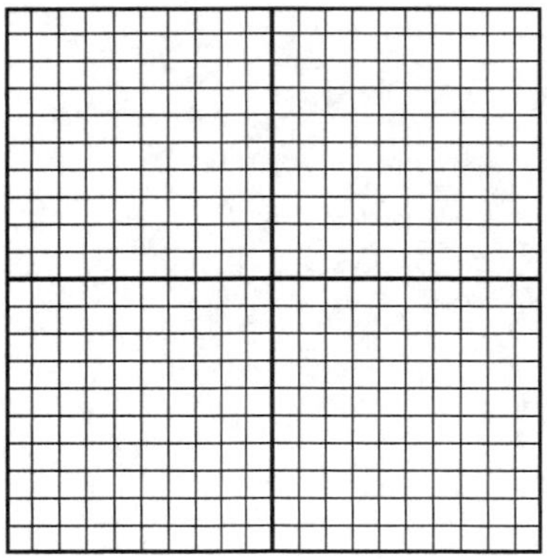

8. $2x - 5y = 0$

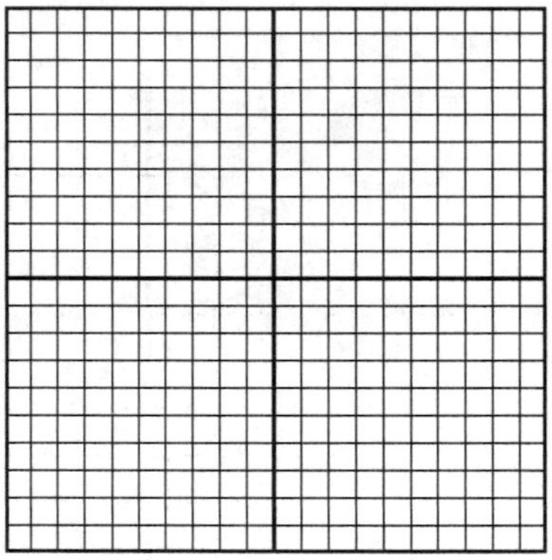

9. $4x - 3y = {}^-2$

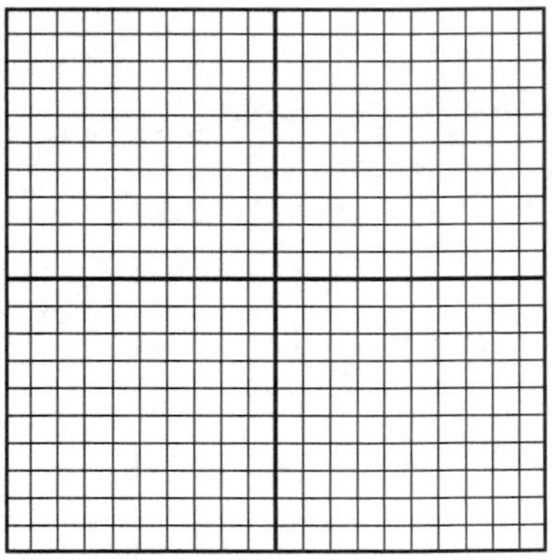

10. $3x + 2y = 6$

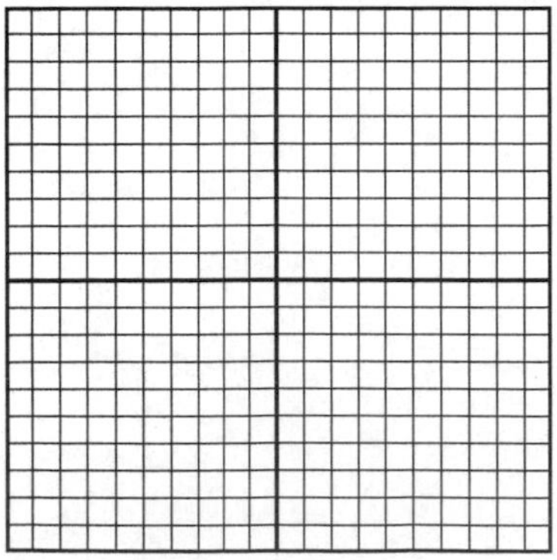

11. $5x - 7y = 12$

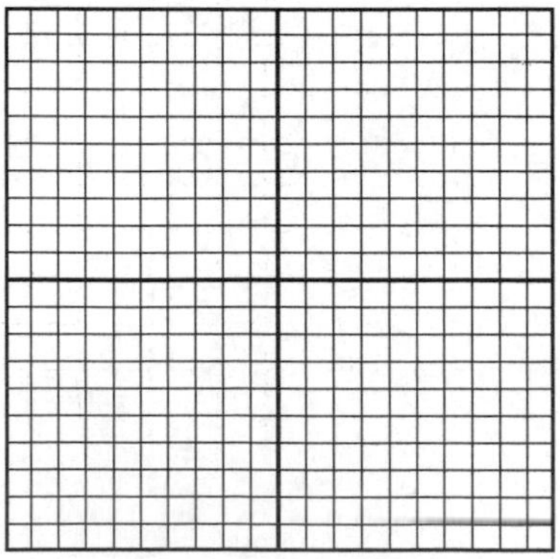

12. $8x + 10y = 50$

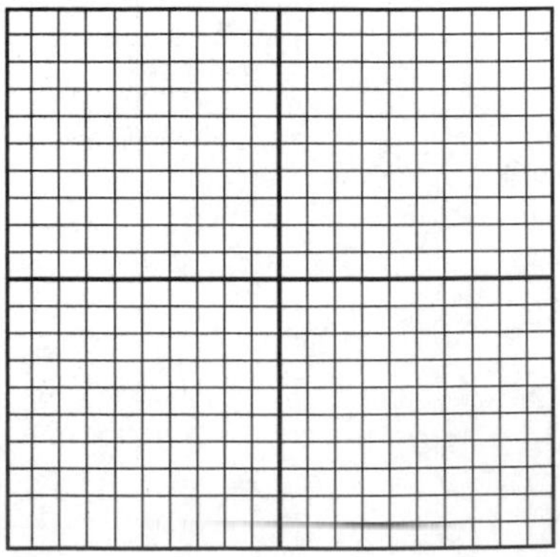

Name________________________________ 8.F.A.1, HSA-REI.D.10, HSA-REI.D.11, HSF-LE.A.2

Graphing Using a Table of Values

Complete the chart for each equation and graph. Charts may vary.

Example: $y = 2x$

x	y
$^-2$	$^-4$
0	0
2	4

1. $y = {}^-3x + 6$

x	y

2. $y = x - 2$

x	y

3. $y + 2x = {}^-2$

x	y

4. $2y + 6x = 5$

x	y

5. $y = \frac{1}{3}x + \frac{3}{2}$

x	y

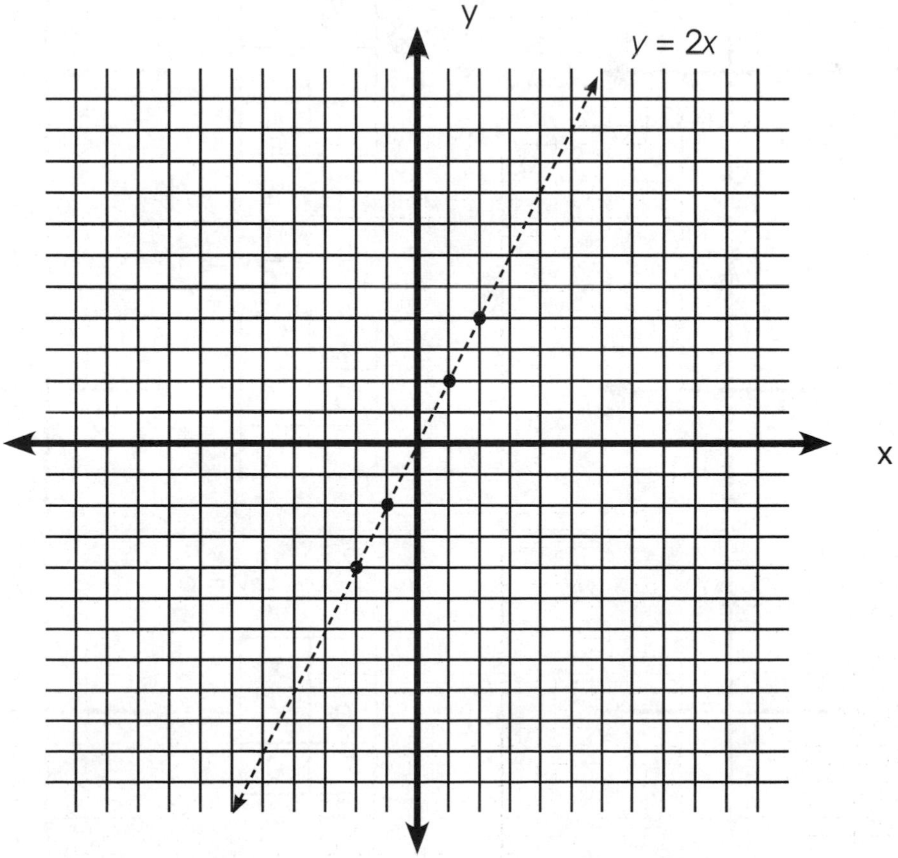

Name____________________ 8.F.A.1, HSA-REI.D.10, HSA-REI.D.11, HSF-LE.A.2

More x and y Intercepts

Graph each linear equation using the *x*- and *y*-intercepts. Add units to both axes.

Example: $2x + 4y = 8$

x	y
0	2
4	0

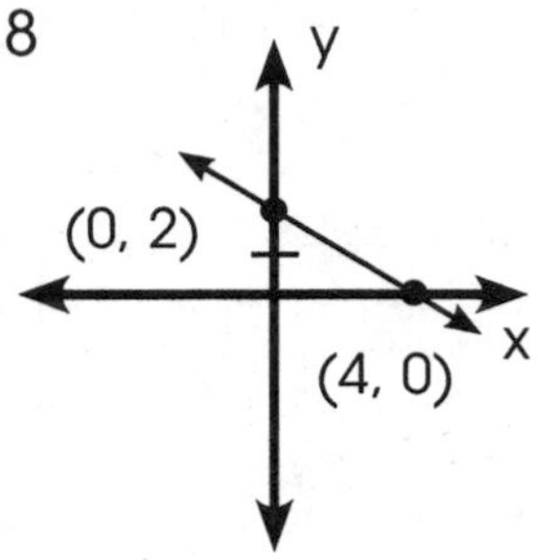

1. $3x - 2y = 12$

x	y
0	
	0

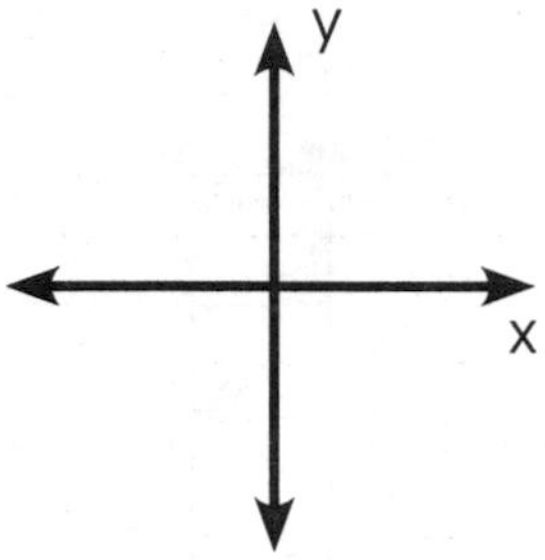

2. $-x + y = 5$

x	y
0	
	0

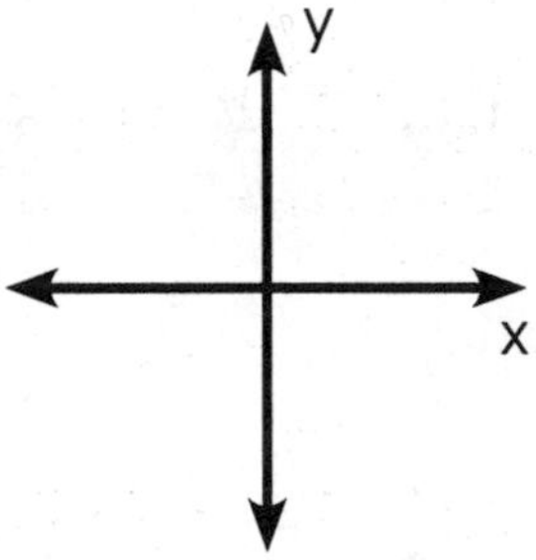

3. $5x - y = 10$

x	y
0	
	0

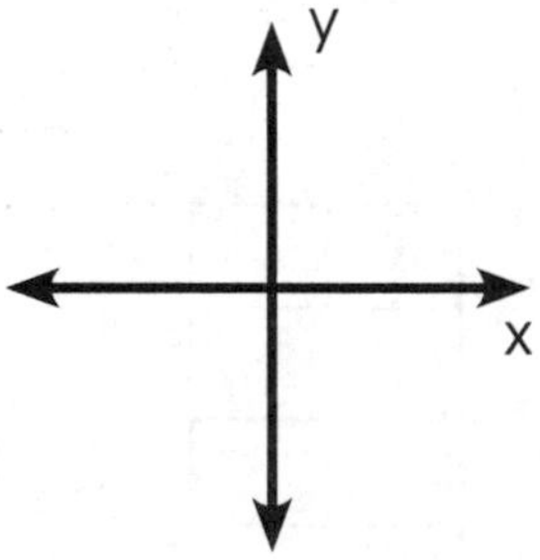

4. $\frac{1}{2}x + y = 5$

x	y
0	
	0

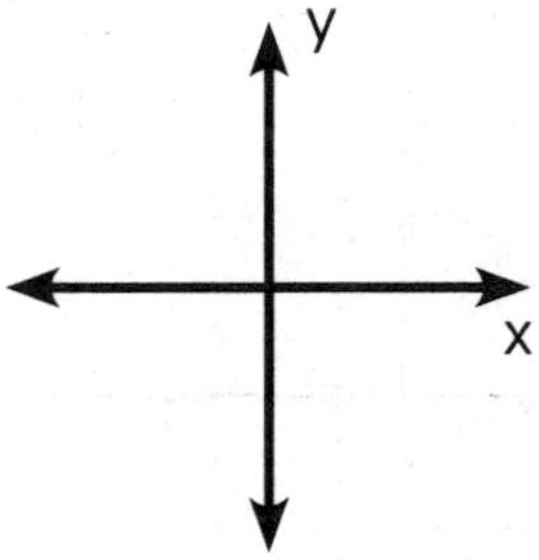

5. $10x - 2y = {}^{-}10$

x	y
0	
	0

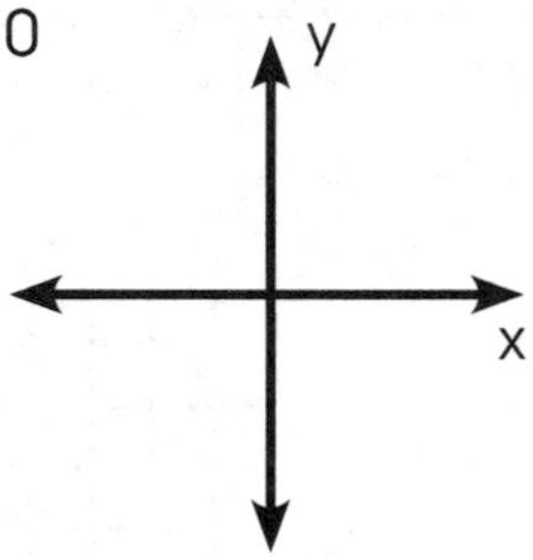

6. $x - \frac{1}{3}y = 2$

x	y
0	
	0

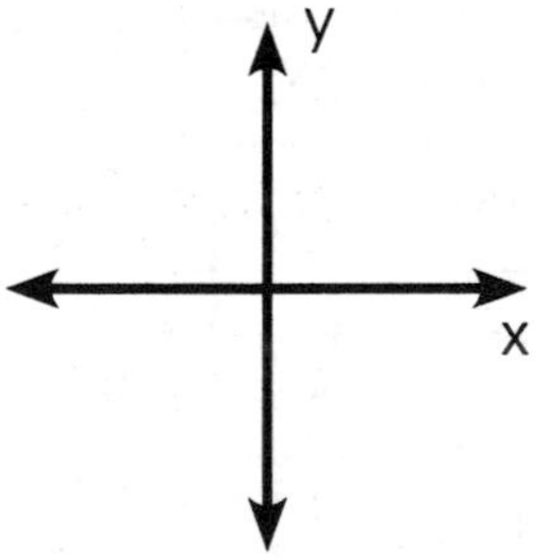

7. $\frac{1}{3}x + \frac{1}{2}y = {}^{-}1$

x	y
0	
	0

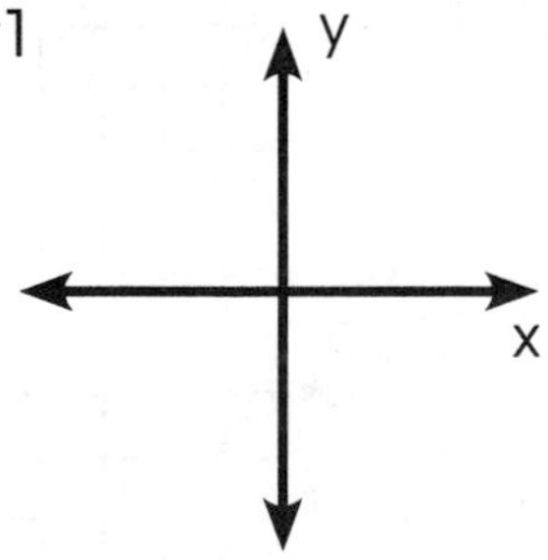

8. $8x - \frac{1}{2}y = 4$

x	y
0	
	0

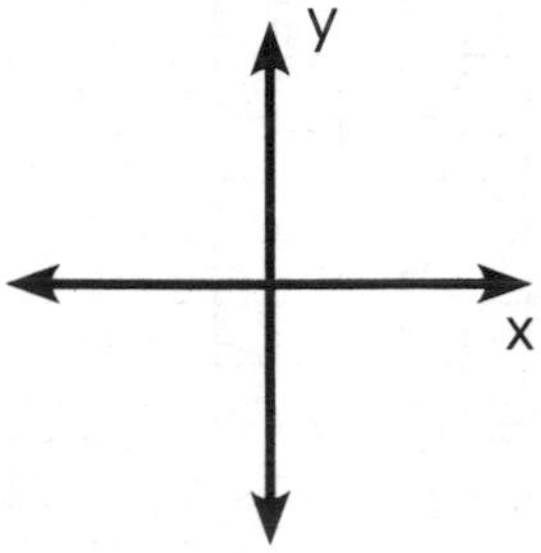

9. $\frac{2}{3}x + \frac{1}{2}y = 6$

x	y
0	
	0

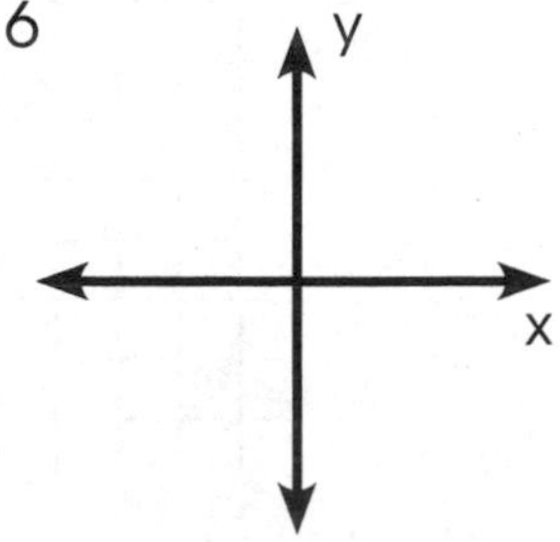

Name________________________________

8.EE.B.5, HSA-REI.D.10

Finding Slope Using a Graph

Find the slope of the lines passing through the given points.

$$\text{slope} = \frac{\text{change in } y}{\text{change in } x}$$

Choose any 2 points to count the change.

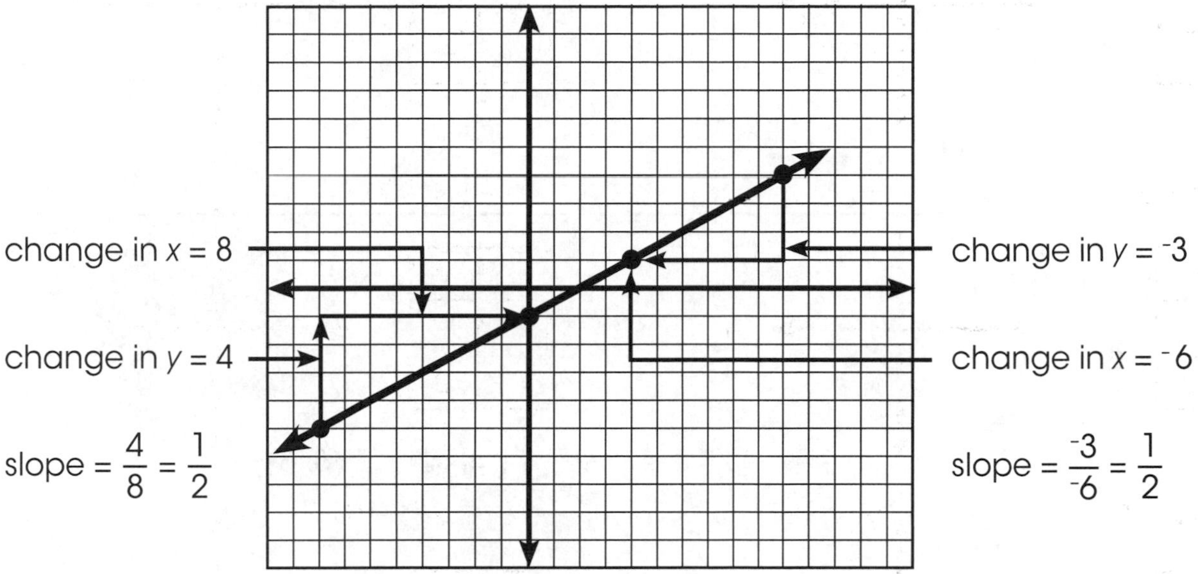

slope is $\frac{1}{2}$

1.

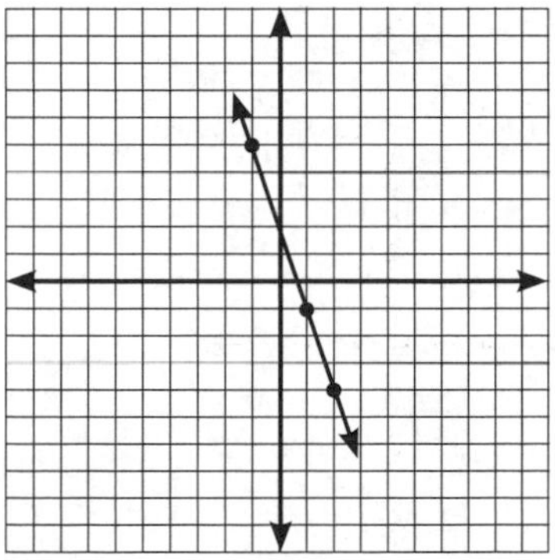

2.

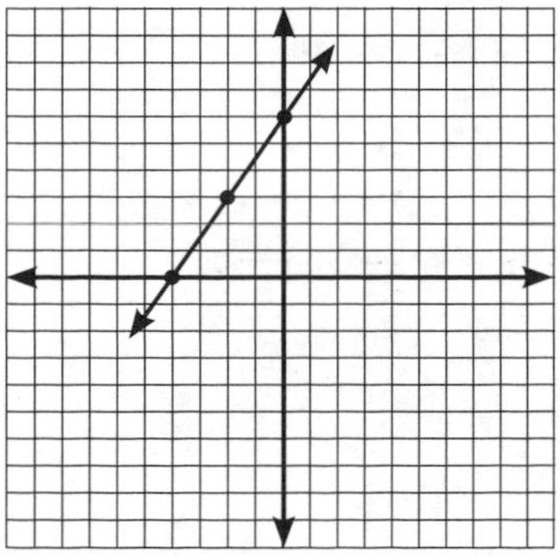

3.

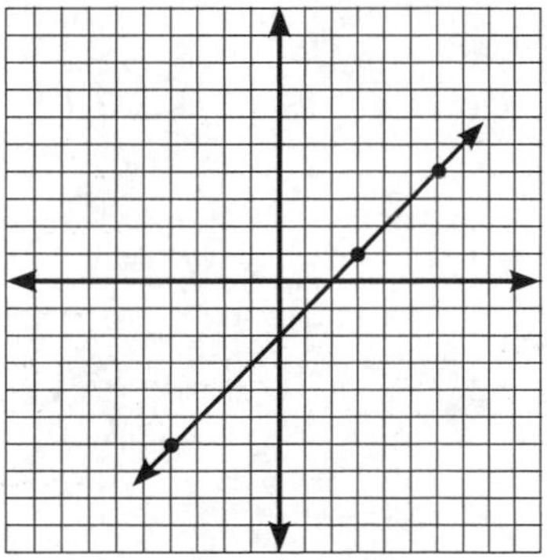

4.

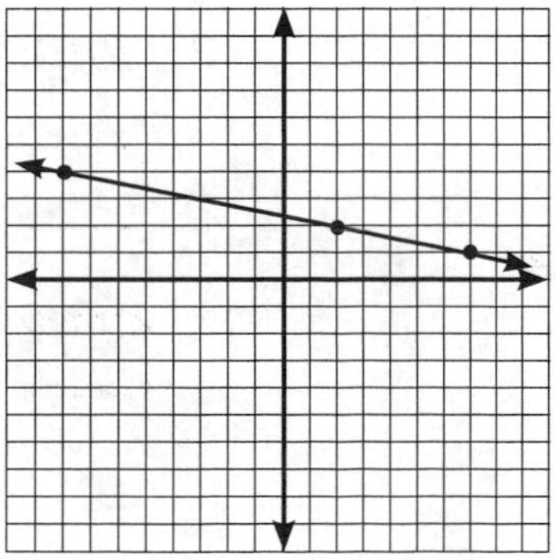

5.

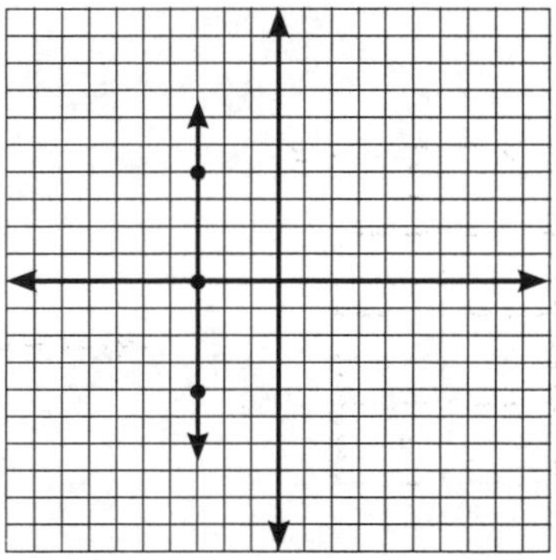

6.

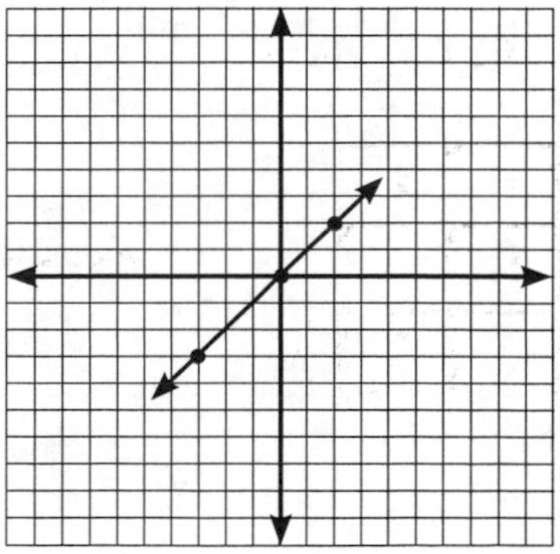

Name______________________________ 8.EE.B.5, HSA-REI.D.11

Finding Slope Using the Slope Formula

$$\text{slope} = \frac{\text{difference in } y\text{-values}}{\text{difference in } x\text{-values}}$$

P (5, 3) R (⁻1, 1)

$$\text{Slope of PR} = \frac{3 - 1}{5 - (^-1)} = \frac{2}{6} = \frac{1}{3}$$

$$\text{slope is } \frac{1}{3}$$

1. A (⁻3, 1) D (4, 5)

2. C (2, 6) F (3, 5)

3. B (0, 8) G (3, 2)

4. J (⁻6, ⁻3) K (⁻4, 5)

5. P (9, 4) M (7, 3)

6. Q (0, ⁻4) R (1, ⁻6)

7. L (⁻2, 6) N (2, ⁻3)

8. S (⁻1, ⁻3) X (2, ⁻6)

9. T (⁻4, ⁻4) Z (6, 3)

10. V $(\frac{3}{4}, \frac{3}{2})$ W $(\frac{11}{4}, \frac{5}{2})$

11. U (2, 3) A (⁻2, 3)

12. C (4, ⁻1) D (⁻2, 2)

13. Z (3, 5) H (5, 10)

14. J (⁻2, ⁻3) K (13, 7)

Name____ 8.EE.B.5, HSA-REI.D.10

Graphing Linear Equations Using Slope

Graph the line that contains the given point and has the given slope.

$(2, {}^-1), \frac{2}{3}$

a. Plot point.

b. Locate other points by moving up 2 units and to the right 3 units.

c. Connect the points with a line.

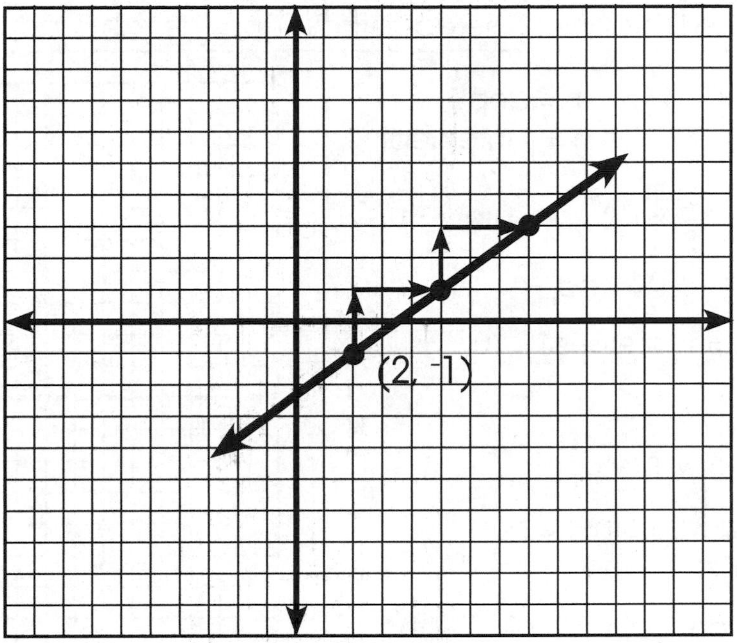

1. $({}^-5, {}^-2), -\frac{1}{2}$

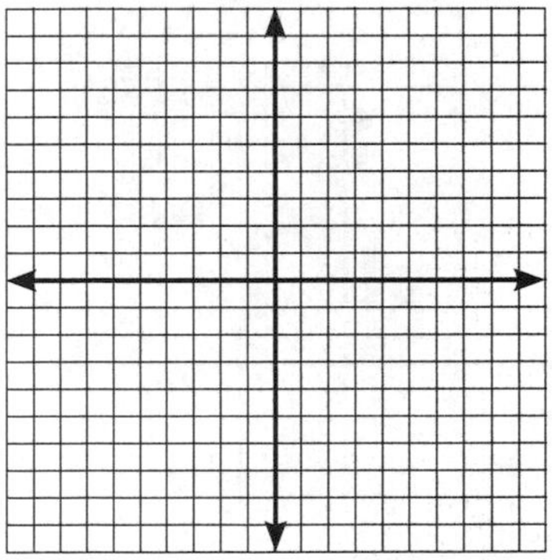

2. $(4, 2), 3$ (**Note:** $3 = \frac{3}{1}$)

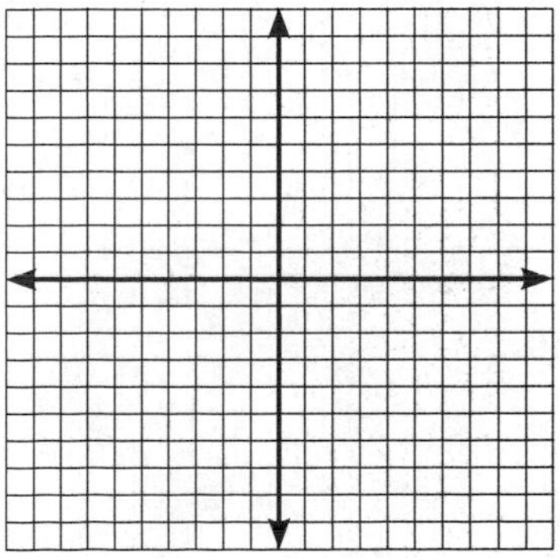

3. $(2, {}^-3), \frac{3}{4}$

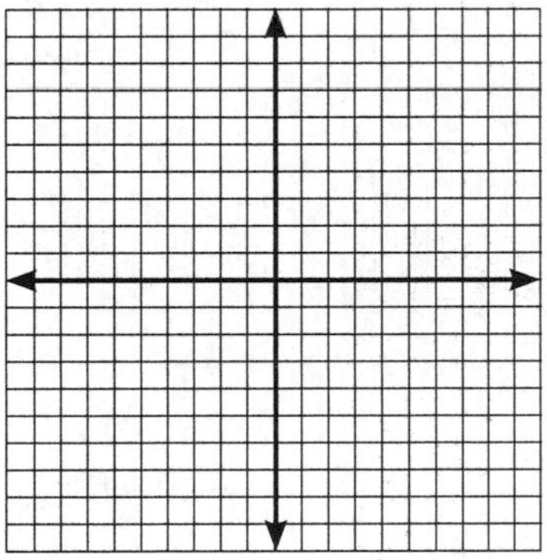

4. $(0, 2), -\frac{5}{2}$

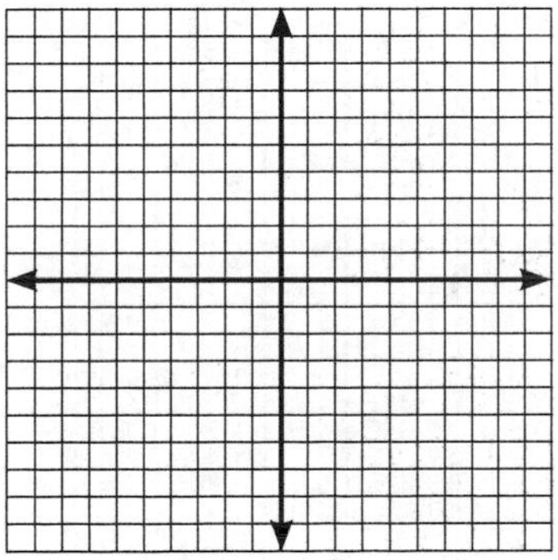

5. $({}^-1, {}^-4), -\frac{1}{4}$

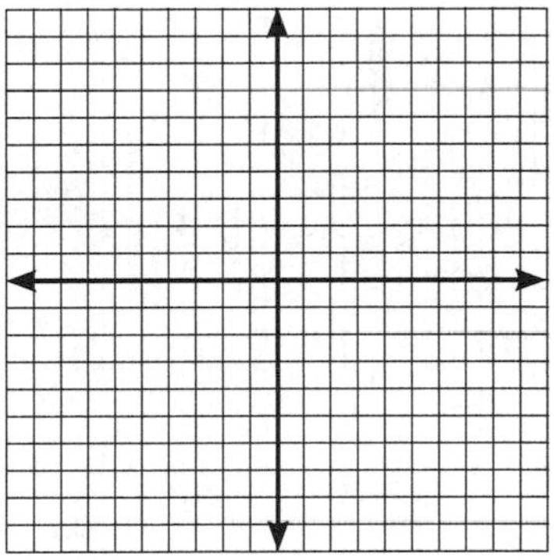

6. $(3, {}^-2), {}^-2$

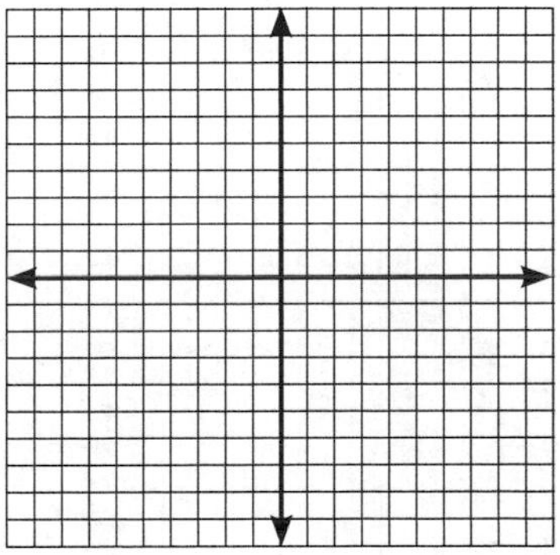

Name________________________________ 8.EE.B.5, HSA-REI.D.10

Graphing Linear Equations Using *y*-Intercept and Slope

Graph the lines given the equation using the y-intercept and slope.

$y = mx + b$ m = slope
b = y-intercept

$y = \frac{2}{3}x + 2$

m = slope = $\frac{2}{3}$

b = y-intercept = (0, 2)

a. Plot y-intercept.

b. Locate other points by using slope.

c. Connect the points with a line.

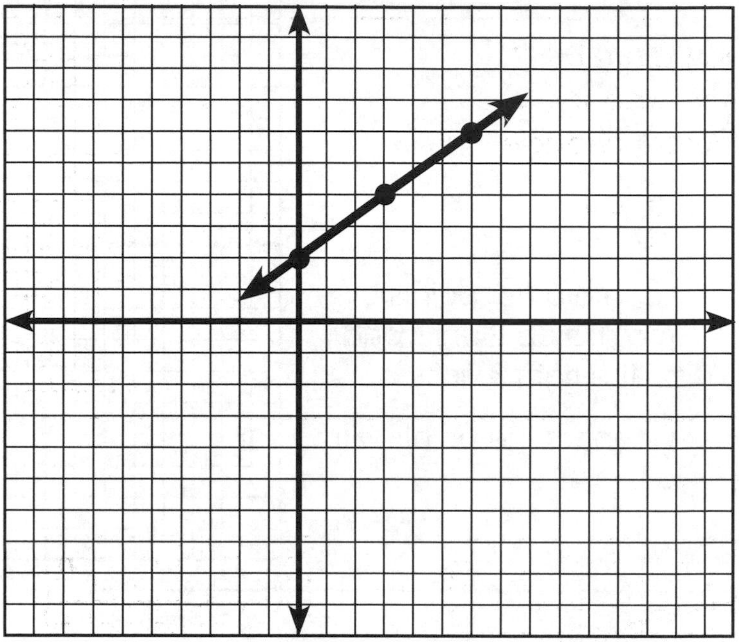

1. $y = \frac{1}{2}x - 1$

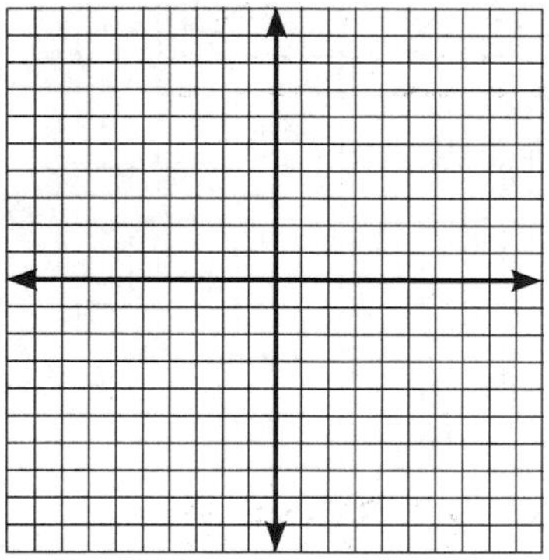

2. $y = 2x + 5$

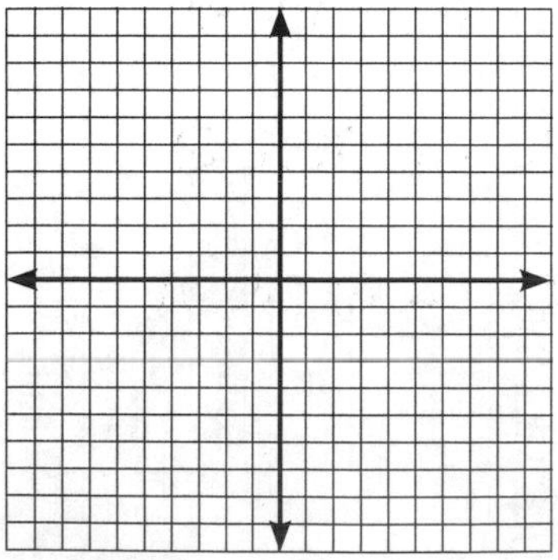

3. $y = -\frac{1}{3}x + 2$

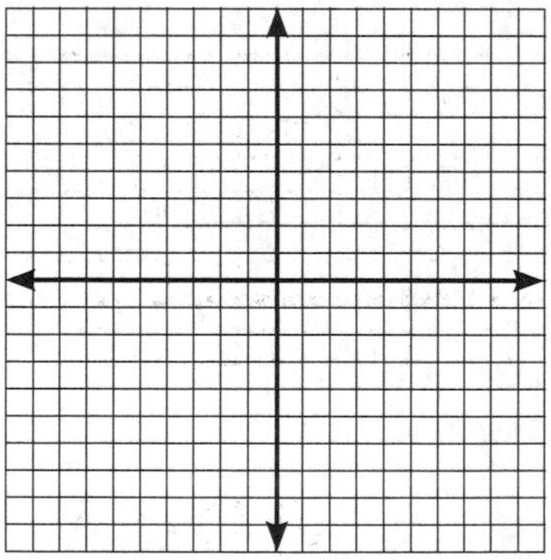

4. $y = {}^{-}3x - 1$

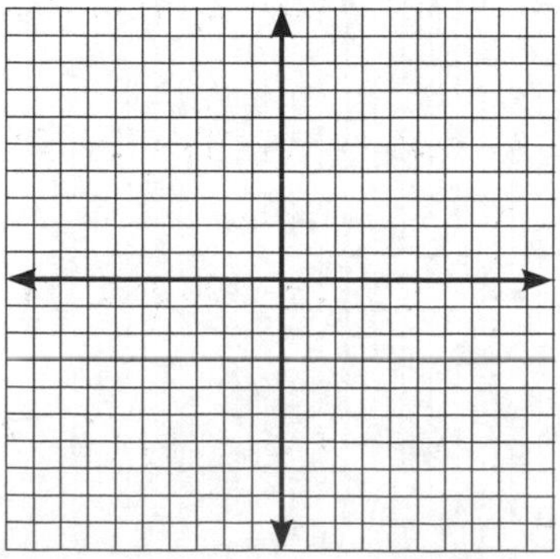

5. $y = -\frac{2}{3}x + 4$

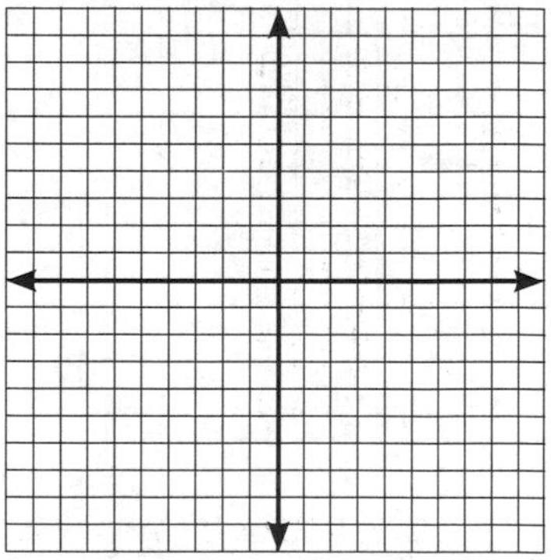

6. $y = x + 3$

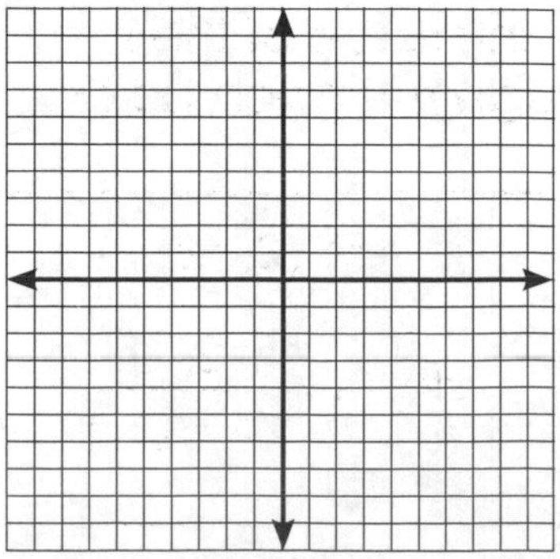

Name____ 8.EE.B.5, HSA-REI.D.11

Finding the Slope of a Line

I. Slope = $\dfrac{\text{vertical change}}{\text{horizontal change}}$

Identify the slope of the line using the graph.

Using points 1 & 2
vertical change = $^-3$
horizontal change = $^-1$

slope = $\dfrac{^-3}{1} = 3$

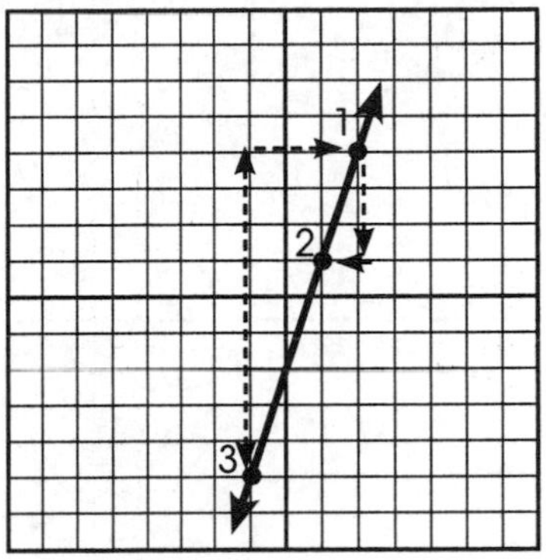

Find the slope.

1.

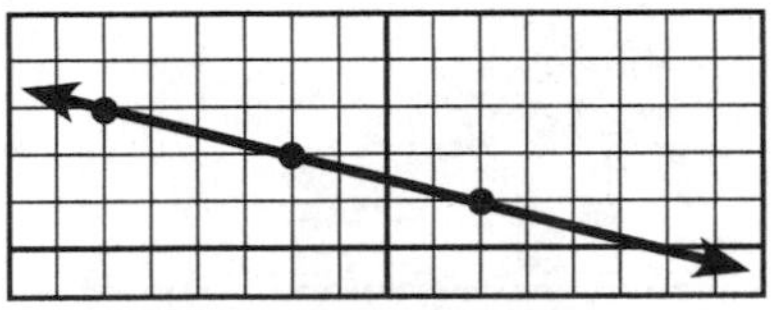

2.

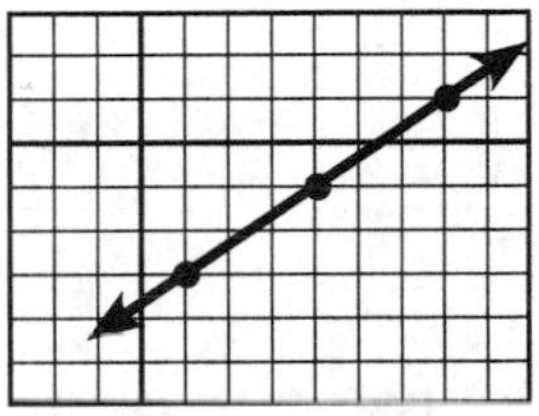

3.

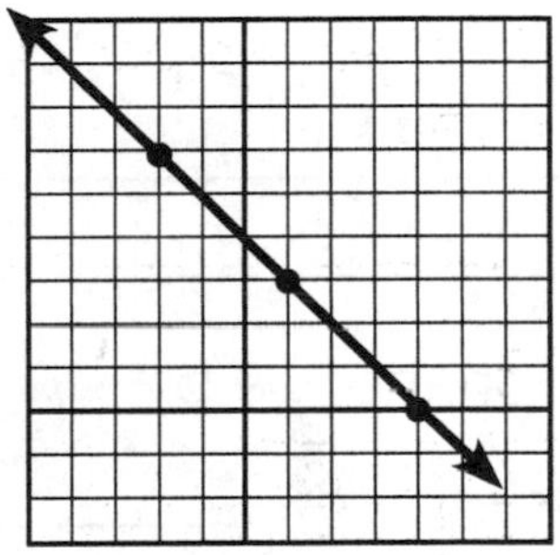

4.

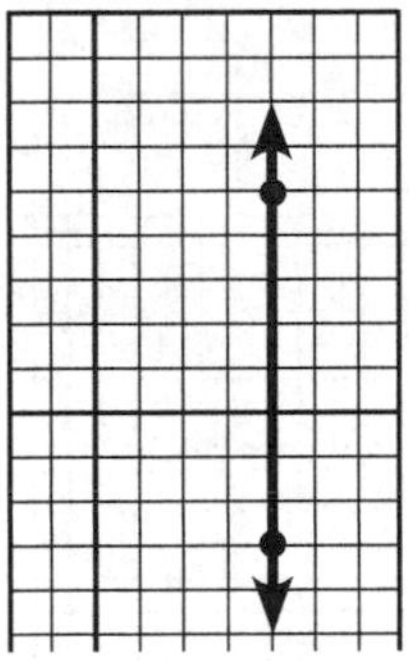

5.

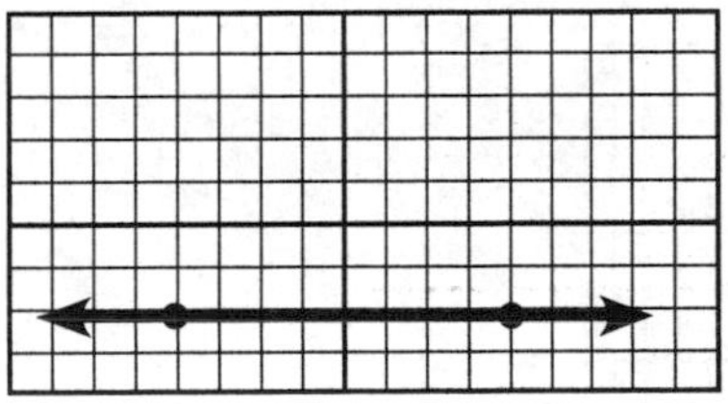

II. Slope = $\dfrac{\text{change in } y\text{-values}}{\text{change in } x\text{-values}} = \dfrac{y_2 - y_1}{x_2 - x_1}$

Find the slope of the line passing through the given points.

$(^-1, 5)\ (3, ^-2)$

slope = $\dfrac{^-2 - 5}{3 - (^-1)} = \dfrac{^-7}{4}$

6. $(0, 0)\ (3, 5)$

7. $(5, ^-2)\ (^-7, 4)$

8. $(^-6, 3)\ (^-2, ^-9)$

9. $(6, ^-9)\ (^-4, 3)$

10. $(^-3, ^-11)\ (2, ^-7)$

11. $(7, 3)\ (^-8, 3)$

12. $(0, 0)\ (4, ^-3)$

13. $(^-2, ^-3)\ (2, 5)$

14. $(^-4, 8)\ (^-4, ^-3)$

Name________________________________

8.EE.B.5, HSA-REI.D.10

Slope-Intercept Form

I. Solve for y.

$$4x + y = 3$$
$$4x - 4x + y = {}^-4x + 3$$
$$y = {}^-4x + 3$$

1. $x + y = 3$

2. $2x - y = 7$

3. ${}^-6 + 2y = 10x$

4. $3y - 6x + 12 = 0$

II. Solve for y; state the m and y_0.

$$9x - 3y = -6$$
$$y = 3x + 2$$
$$m = \frac{3}{1}$$
$$y_0 = 2$$

5. $2y - 6x = 2$

6. $y - 4x = {}^-3$

7. $4y = 5x + 12$

8. $2x - 3y = 5$

III. Graph the line by 1.) solving for y 2.) using m and y_0.

9. $4x + y = {}^-8$

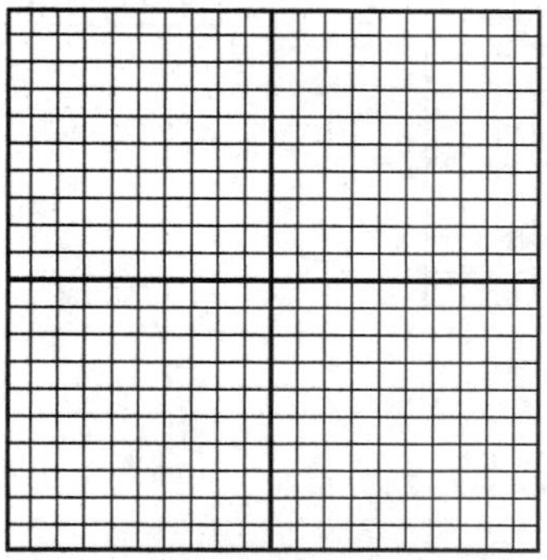

10. $y - 3x = {}^-9$

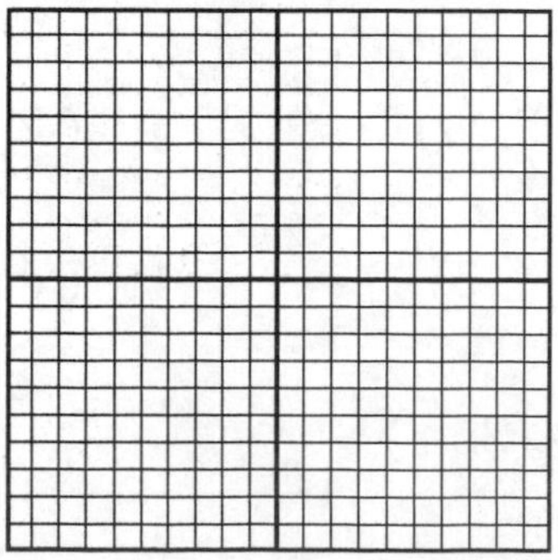

11. $2x - 4y = {}^-16$

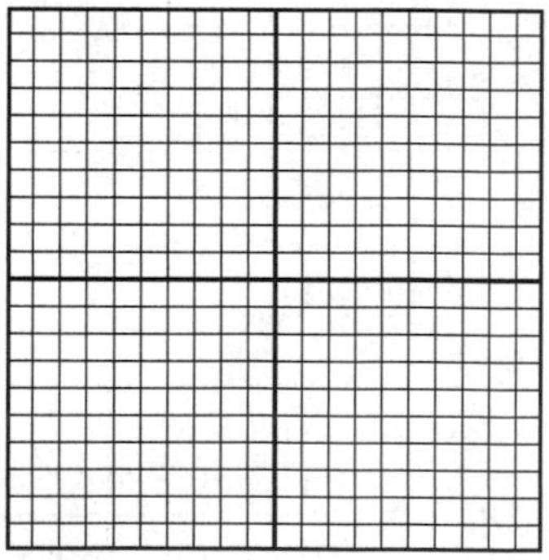

12. $3x + 3y + 4 = 0$

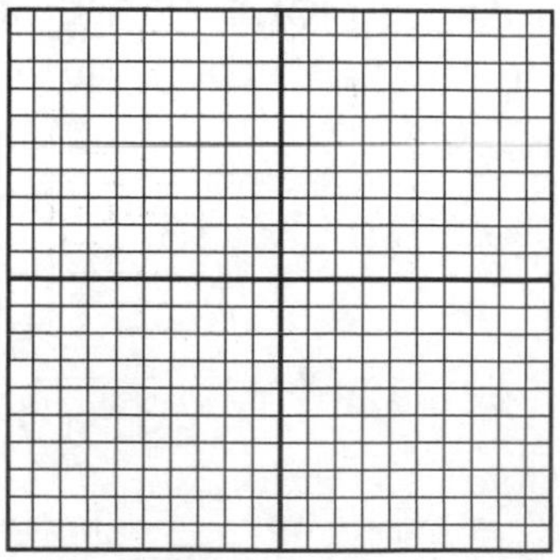

Name________________________________ 8.EE.B.5, HSA-REI.D.10

Graphing Using Slope-Intercept Form

Graph from the slope-intercept form: $y = mx + b$ m = slope b = y-intercept

$y = 4x + 2$

1) Plot y-intercept.

   $b = 2$ ⟹ (0, 2)

2) Find other points using slope.

   $m = 4$ ⟹ $\frac{4}{1}$ or $\frac{-4}{-1}$

3) Connect points.

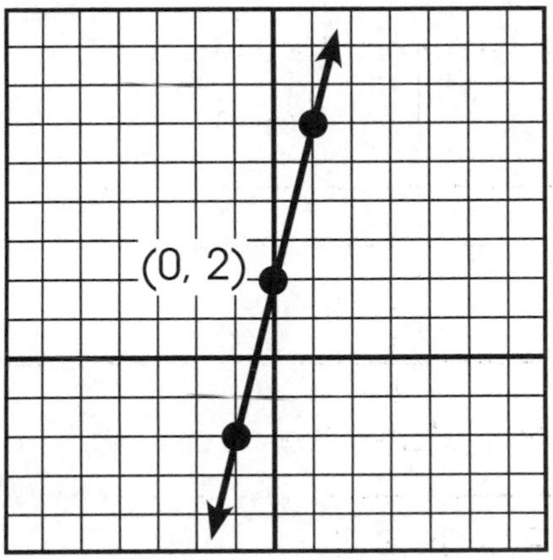

1. $y = 2x - 4$

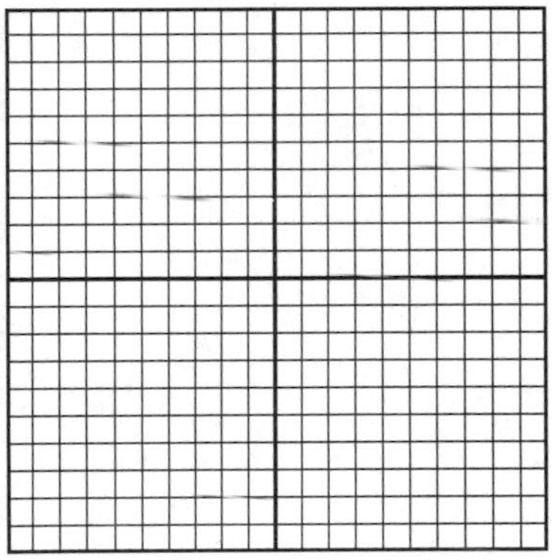

2. $3x - y = 7$

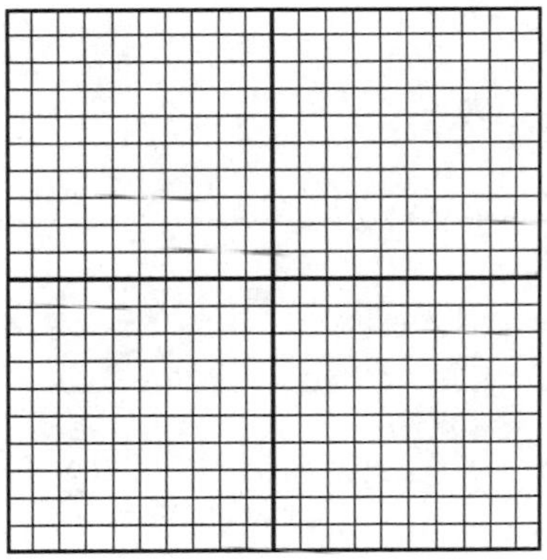

3. $2x + 3y = 6$

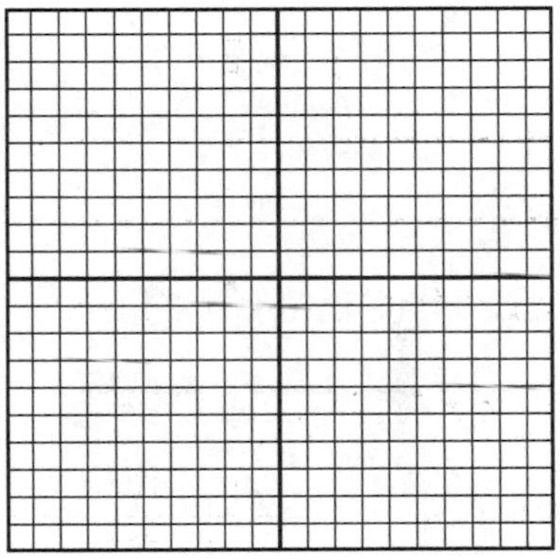

4. $y = -\frac{2}{3}x + 1$

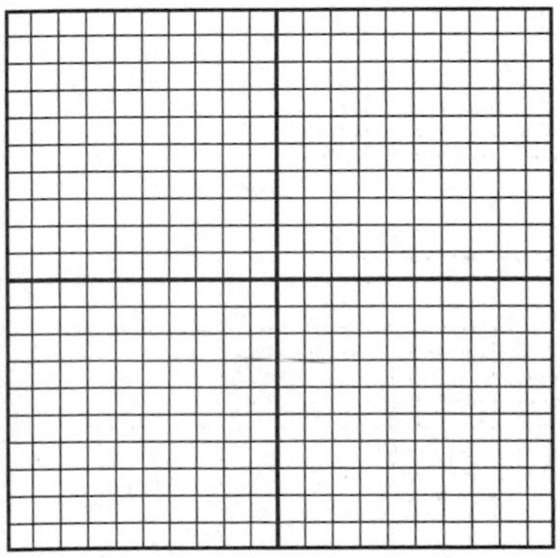

5. $x - 4y + 8 = 0$

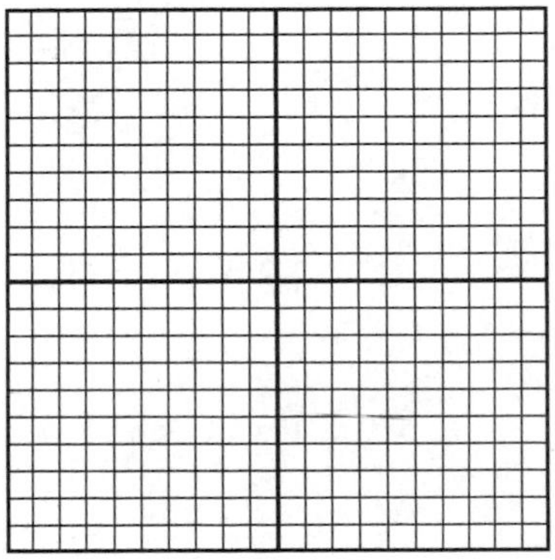

6. $6x - 5y = 15$

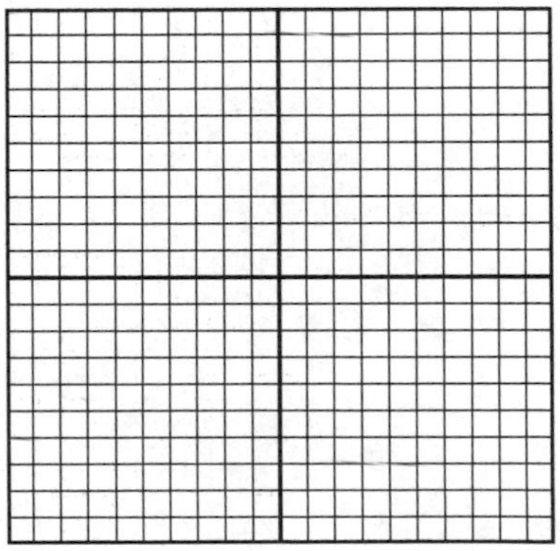

Name________________________________ 8.EE.B.5, HSA-CED.A.2, HSA-REI.D.11

Equations in Standard Form: Ax + By = C

I. Put in standard form.

$$y = -\frac{2}{5}x + 3$$

$$^{-}5y = 2x - 15$$

$$^{-}2x - 5y = {}^{-}15$$

$$2x + 5y = 15$$

1. $y = -\frac{3}{4}x + 2$
2. $y = \frac{1}{2}x - 2$
3. $y = 3x + 6$
4. $y = -x - 5$
5. $y = \frac{3}{4}x + \frac{1}{2}$
6. $y = -\frac{1}{4}x + 8$

II. Find the equation of a line in standard form using the slope-intercept form.

$m = -\frac{3}{4}$ $y_0 = 2$

$$y = -\frac{3}{4}x + 2$$

$$\frac{3}{4}x + y = 2$$

$$3x + 4y = 8$$

7. $m = 3$ $y_0 = -\frac{1}{2}$
8. $m = \frac{5}{4}$ $y_0 = 2$
9. $m = -\frac{2}{3}$ $y_0 = \frac{3}{5}$
10. $m = 4$ $y_0 = {}^{-}3$
11. $m = \frac{3}{4}$ $y_0 = \frac{1}{2}$
12. $m = \frac{7}{2}$ $y_0 = -\frac{3}{4}$
13. $m = 0$ $y_0 = {}^{-}3$

Name________________________________ 8.EE.B.5, HSA-CED.A.2, HSA-REI.D.11, HSF-LE.A.2

Equations in Standard Form: Using Point-Slope Formula

III. Find the equation of the line in standard form using the point-slope formula.

$m = 3$ (1, 2)

Let (x, y) be any other point on the line.

Use slope formula:

$$m = \frac{y_2 - y_1}{x_2 - x_1}$$

or point-slope formula

$$y_2 - y_1 = m\,(x_2 - x_1)$$

$$3 = \frac{y - 2}{x - 1}$$

$$3\,(x - 1) = y - 2$$

$$3x - 3 = y - 2$$

$$3x = y + 1$$

$$3x - y = 1$$

1. $m = {}^{-}3$, $(4, 5)$

2. $m = {}^{-}2$, $(1, 3)$

3. $m = 0$, $(4, {}^{-}6)$

4. $m = \frac{3}{4}$, $(1, 0)$

5. $m =$ no slope, $({}^{-}3, \frac{3}{4})$

6. $m = {}^{-}1$, $({}^{-}1, 4)$

7. $m = -\frac{1}{2}$, $(6, {}^{-}3)$

8. $m = 1$, $(1, {}^{-}4)$

9. $m = \frac{1}{4}$, $({}^{-}4, 3)$

10. $m = \frac{1}{3}$, $({}^{-}3, {}^{-}2)$

11. $m = \frac{2}{3}$, $(1, 1)$

12. $m = 0$, $(7, {}^{-}4)$

13. $m = -\frac{2}{1}$, $({}^{-}2, {}^{-}7)$

14. $m = \frac{5}{1}$, $({}^{-}2, 0)$

Name____________________________________

8.EE.B.5, 8.F.A.1, HSA-CED.A.2, HSF-LE.A.2

Equations in Standard Form: Given Two Points

IV. Find the equation of the line in standard form using 1) slope and then 2) point-slope formula.

$$(^-3, 4)\ (4, 7)$$

$$m = \frac{7 - 4}{4 - (^-3)} = \frac{3}{7}$$

$$\frac{3}{7} = \frac{y - 7}{x - 4}$$

$$y - {}^-4 = \frac{3}{7}(x + 3)$$

$$7y - 28 = 3x + 9$$

$$3x - 7y = {}^-37$$

1. (2, 1) (4, 0)

2. (5, 2) (2, $^-$1)

3. (4, $^-$3) (0, 3)

4. ($^-$2, $^-$3) ($^-$1, 2)

5. (0, 0) ($^-$1, $^-$2)

6. (6, $^-$3) ($^-$2, $^-$3)

7. (2, 3) ($^-$1, 5)

8. (4, 8) (4, $^-$2)

9. (5, 8) (3, 2)

10. ($^-$2, 5) (3, $^-$10)

11. (0, 2) ($^-$4, 2)

12. ($^-$1, $^-$1) (0, $^-$4)

13. ($^-$3, 6) ($^-$3, 2)

14. ($^-$6, 6) (3, 3)

Name________________________________ (8.EE.B.5, 8.F.A.1, HSA-CED.A.2, HSF-LE.A.2)

Slope Challenge

Find the slope of the line passing through two points using the formula:

$$\text{slope} = m = \frac{(y_2 - y_1)}{(x_2 - x_1)}.$$

Example: (1, 1) (4, 2)

(x_1, y_1) (x_2, y_2)

$$m = \frac{(2 - 1)}{(4 - 1)} = \frac{1}{3}$$

1. (15, ⁻12) (10, ⁻2)
2. (5, ⁻12) (15, ⁻2)
3. (⁻3, 14) (⁻1, 28)
4. (9, 6) (4, 6)
5. (8, 14) (22, ⁻9)
6. (33, 59) (0, 0)
7. (14, 21) (14, 6)
8. (5, 17) (⁻18, 9)
9. (16, ⁻1) (8, 9)
10. (⁻4, 2) (6, 9)
11. (8, 4) (7, ⁻3)
12. (⁻1, ⁻19) (⁻15, 4)
13. (⁻8, ⁻1) (6, 6)
14. Parallel lines have equal slopes. Two pairs of the lines indicated by the given points are parallel. Give the problem numbers of the parallel lines:

    __________ & __________. __________ & __________.

15. Perpendicular lines intersect to form right angles. (If they are not vertical or horizontal, then the product of the slopes equals ⁻1.) Two pairs of the lines indicated by the given points are perpendicular. Give the problem numbers of the perpendicular lines:

    __________ & __________. __________ & __________.

16. What can be said about the relationship of any vertical line to any horizontal line?

    ____

Name____

8.EE.B.5, 8.F.A.1, HSA-REI.D.11

Slope and *y*-Intercept

Find the slope and *y*-intercept from each graph.

$$\text{Slope} = \frac{\text{rise}}{\text{run}} = \frac{(y_2 - y_1)}{(x_2 - x_1)}$$

Hint: Think of rise as "ryse" to remind you that the *y*'s go on top.

Example:

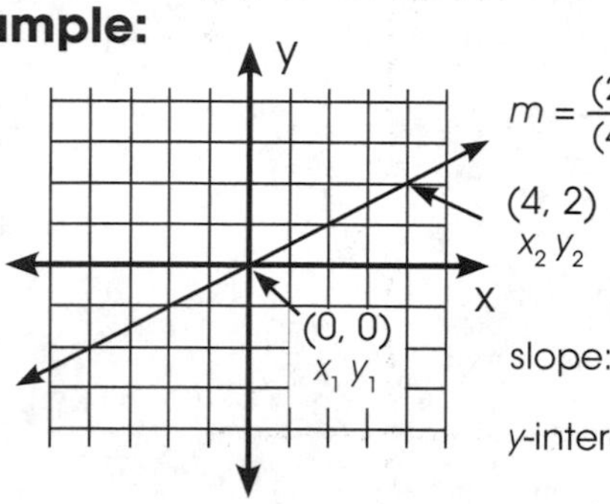

$m = \frac{(2 - 0)}{(4 - 0)} = \frac{2}{4} = \frac{1}{2}$

slope: $\frac{1}{2}$

y-intercept: (0, 0)

1.

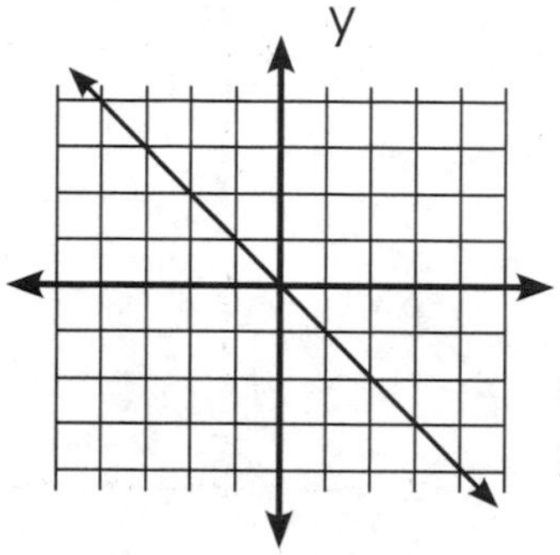

slope:

y-intercept:

2.

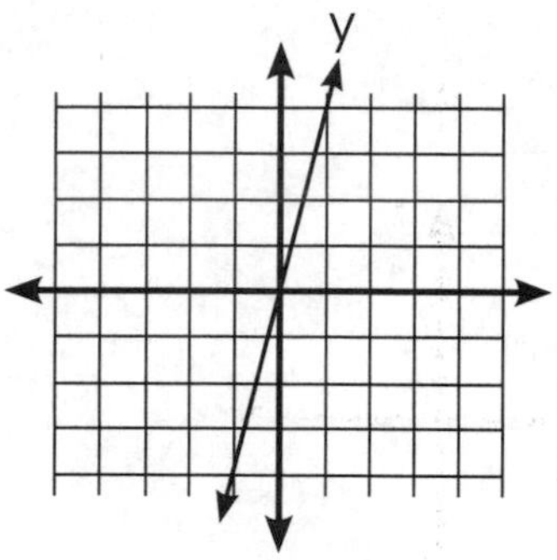

slope:

y-intercept:

3.

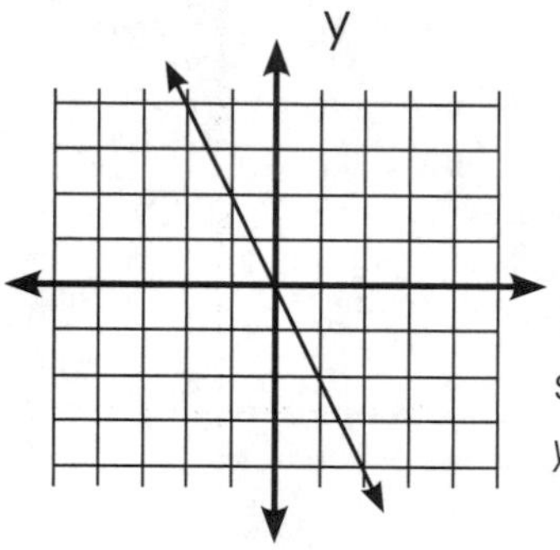

slope:

y-intercept:

4.

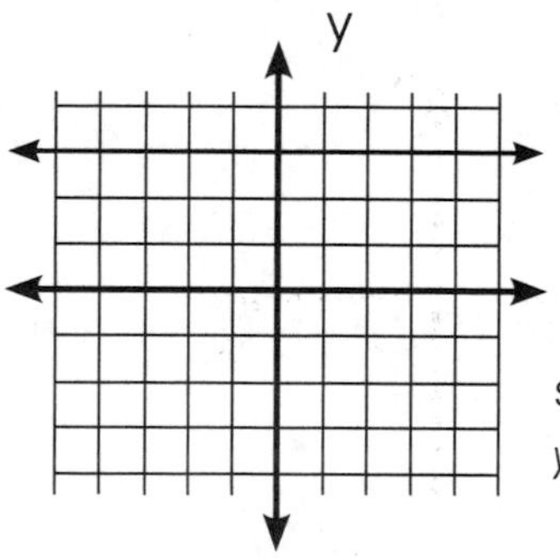

slope:

y-intercept:

5.

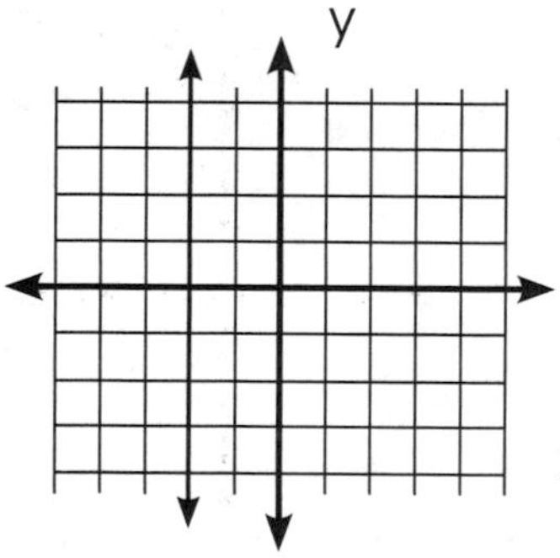

slope:

y-intercept:

6.

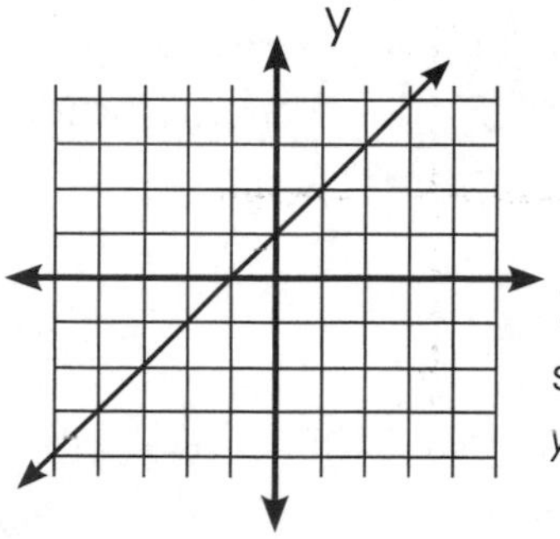

slope:

y-intercept:

7.

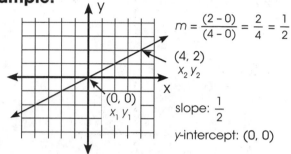

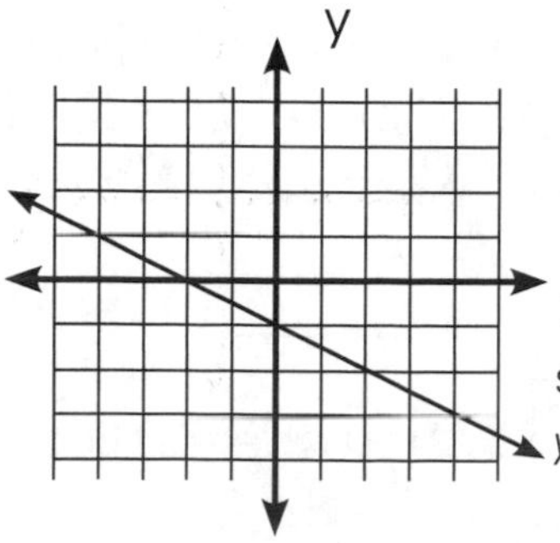

slope:

y-intercept:

Name________________________________ 8.EE.B.5, 8.F.A.1, HSA-REI.D.10

Slope and *y*-Intercept Challenge

Graph each linear equation using the slope and *y*-intercept. Add units to both axes.
$y = mx + b$, where m = slope and b = y-intercept.

Example: $y = 2x + 1$

y – int (0, 1)

$m = \frac{2}{1} = \frac{\text{up } 2}{\text{right } 1}$

Note: Move up 2 and over one from y-intercept.

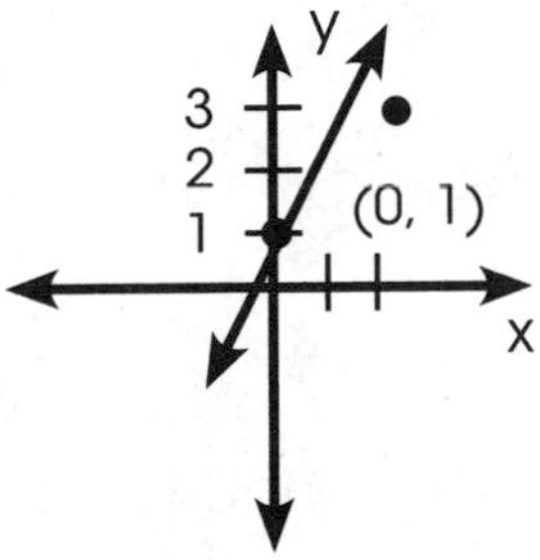

1. $y = -x + 6$

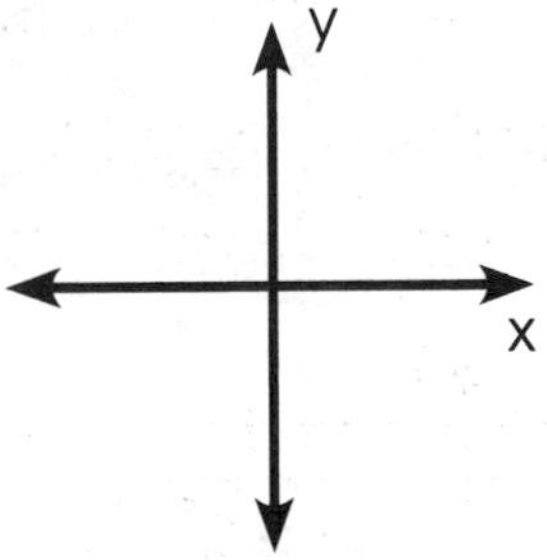

2. $y = \frac{-2}{3}x - 1$

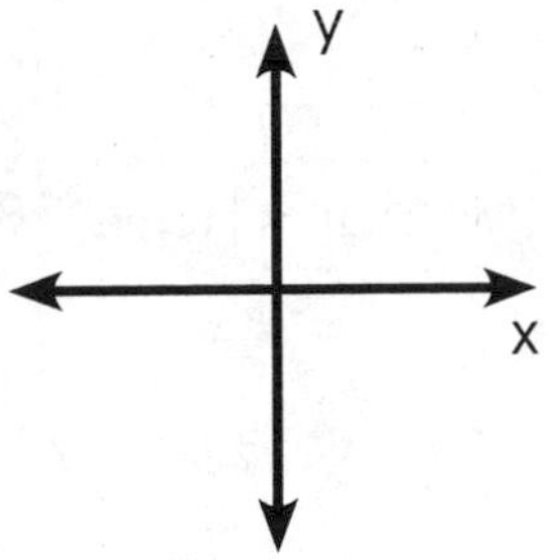

3. $y = 5x + 5$

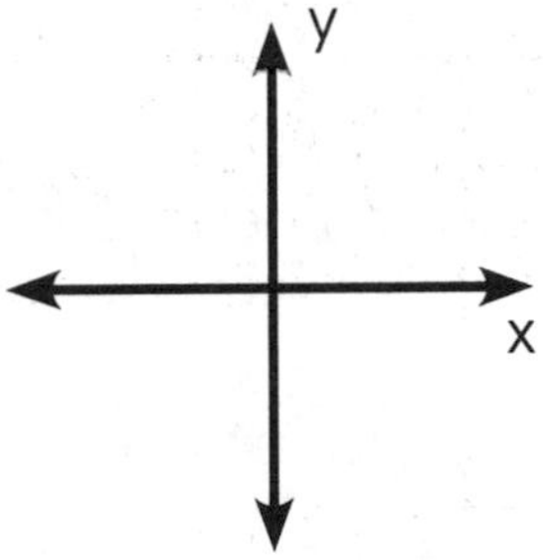

4. $y = 4x$

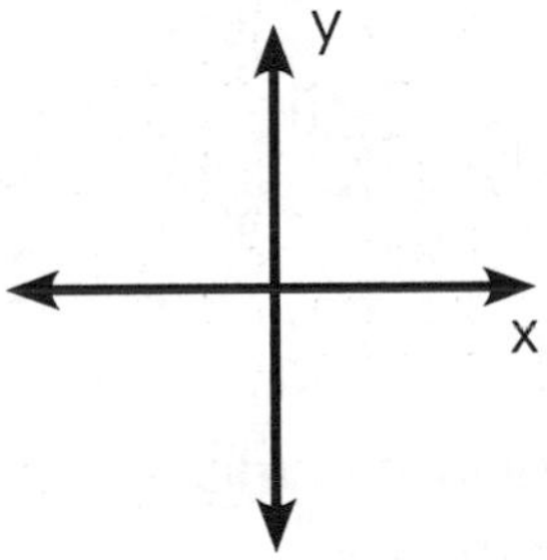

5. $y = 4$

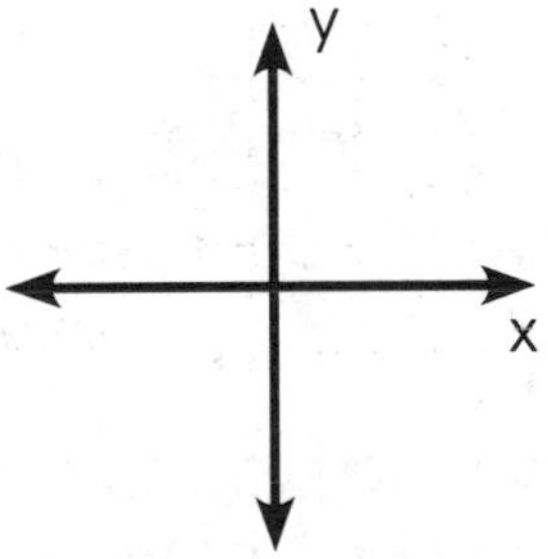

6. $x = -9$

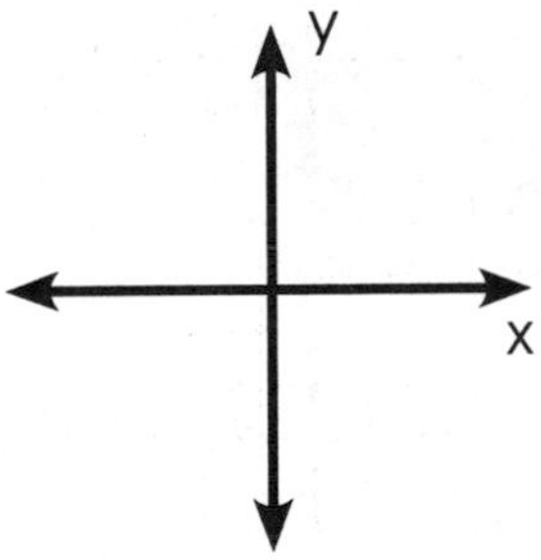

7. $y = -2x - 4$

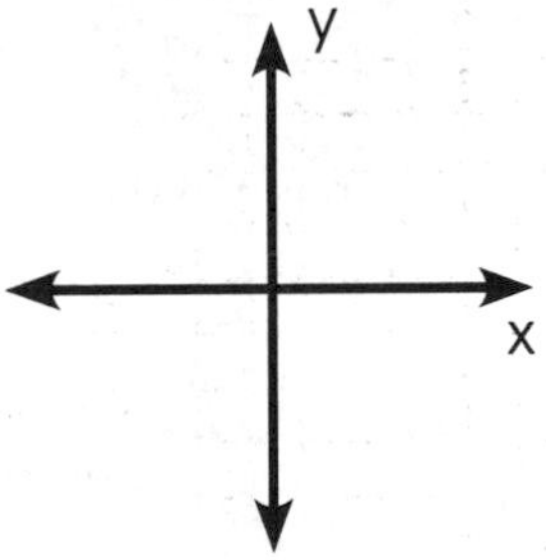

8. $y = \frac{3}{5}x$

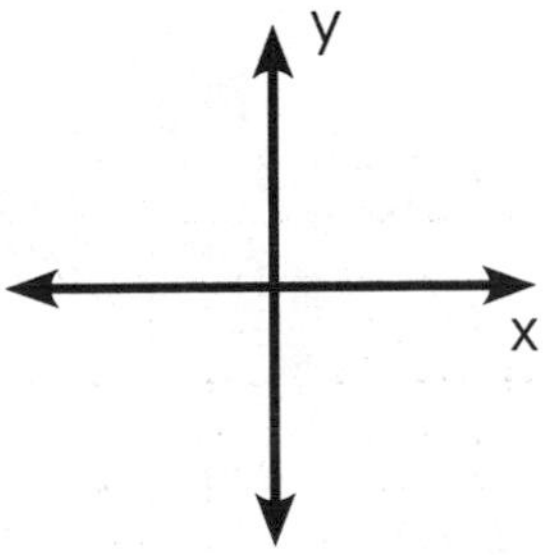

9. $y = \frac{-1}{2}x + 3$

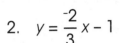

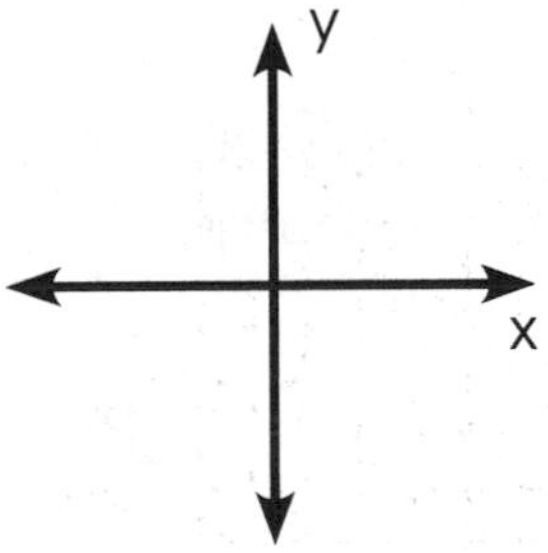

Name________________________________

Just for Fun

We often make mistakes by missing the obvious. Here are some examples. These are not exactly tricky questions. They are rather examples in which the obvious has been overlooked. Our habits and practices lead us to do this often.

1. If it takes 5 minutes to make one cut across a log, how long will it take to cut a 5-foot log into 5 equal pieces?

2. How can two fathers and two sons divide three automobiles among themselves with each receiving one?

3. Some months have 30 days, some have 31. How many have 28 days?

4. If a doctor gave you three pills and told you to take one every half hour, how long would they last?

5. I have two U.S. coins in my hand which total fifty-five cents. One is not a nickel. What are the coins?

6. Two men are playing chess. They played five games and each man won the same number of games with no ties. How is this possible?

7. Why can't a man living in St. Louis be buried in Illinois?

8. If dirt weighs 100 lb. per cubic foot, what is the weight of dirt in a hole three feet square by two feet deep?

9. If you had only one match and entered a dark room in which there was a kerosene lamp, an oil burner, and a wood burning stove, which would you light first?

10. Is there a fourth of July in England?

Name________________________________ 7.EE.B.4b, 8.EE.C.7a, HSA-REI.B.3

Basic Inequalities: Solve and Graph

$$6 < 3\,(1 - s)$$
$$6 < 3 - 3s$$
$$6 - 3 < 3 - 3 - 3s$$
$$\frac{3}{^{-}3} < \frac{^{-}3s}{^{-}3}$$
$$^{-}1 \boxed{>} s$$
$$s < {}^{-}1$$

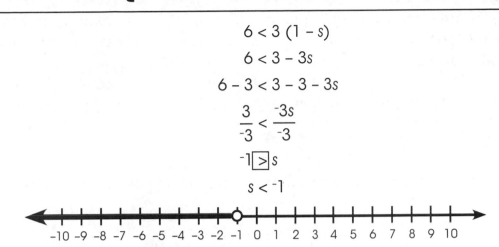

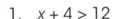

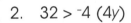

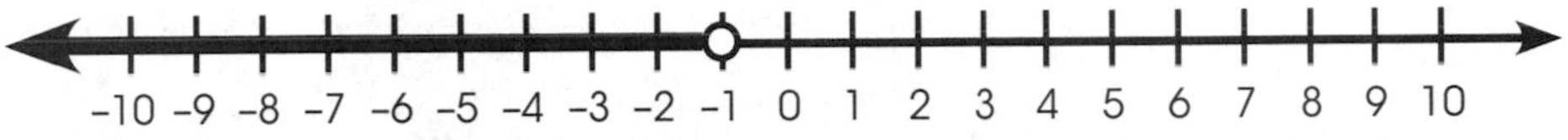

1. $x + 4 > 12$

2. $32 > {}^{-}4\,(4y)$

3. $3y + 1 < 13$

4. $10\frac{1}{2} < 2z + 18\frac{1}{2}$

5. $2 - \frac{n}{3} < {}^{-}1$

6. $^{-}2x - 5 > 6$

7. $^{-}3m + 6\,(m - 2) > 9$

8. $15x - 2 < 3x - 11$

9. $2\,(t + 3) < 3\,(t + 2)$

10. $15x - 2\,(x - 4) > 3$

11. $x - 1.5 < 0.5\,(x + 4)$

12. $^{-}3\,(2m - 8) < 2\,(m + 14)$

13. $2x + 3 < 6x - 1$

14. $3x - 2 \geq 7x - 10$

Name________________________________ 8.EE.C.7a, HSA-REI.B.3

Compound Inequalities: Solve and Graph

$\frac{2x}{2} \geq \frac{^-6}{2}$

$x \geq 3$

$x > {^-3}$ or $x = {^-3}$

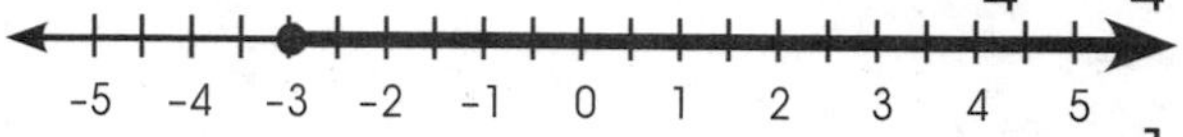

$3 - 4a \leq 5$ or $5a + 1 < {^-4}$

$3 - 3 - 4a \leq 5 - 3$ $\qquad$ $5a + 1 - 1 < {^-4} - 1$

$\frac{^-4a}{^-4} \leq \frac{2}{^-4}$ $\qquad$ $\frac{5a}{5} < \frac{^-5}{5}$

$a \geq -\frac{1}{2}$ or $a < {^-1}$

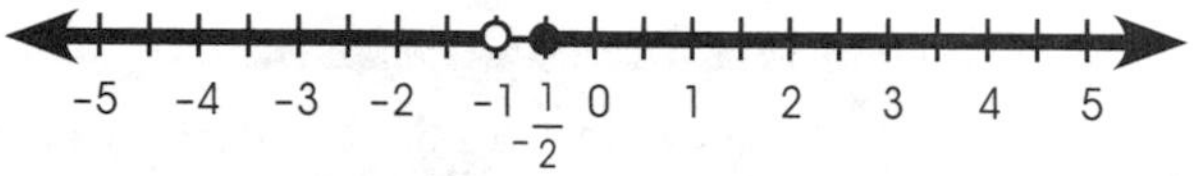

$2 \leq y + 3 < 6$

$2 \leq y + 3$ and $y + 3 < 6 - 3$

$2 - 3 \leq y + 3 - 3$ $\qquad$ $y + 3 - 3 < 6 - 3$

${^-1} \leq y$ and $y < 3$

$y \geq {^-1}$

1. $t \leq {^-1}$ or $t \leq {^-3}$

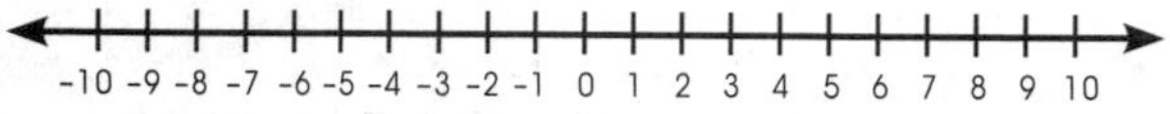

2. ${^-2} \leq x \leq 6$

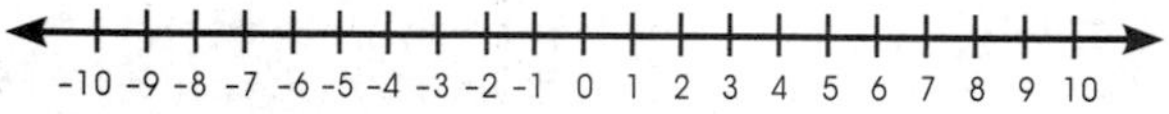

3. $x + 1 \leq {^-3}$ or $x \geq 3$

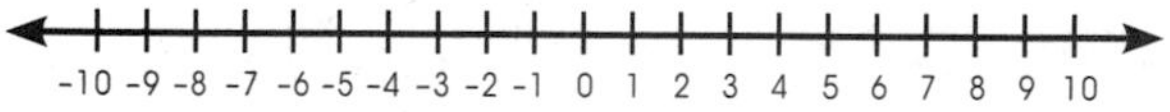

4. ${^-2} < 3t - 2 < 10$

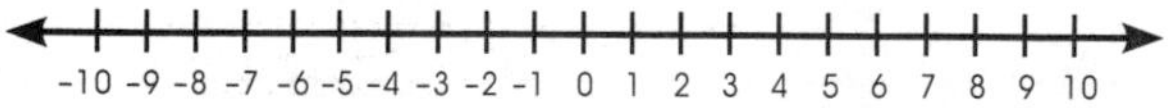

5. $-(x - 2) \geq 3$

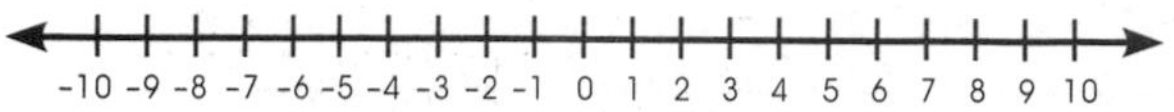

6. $3x - 7 < 11$ or $9x - 4 > x + 4$

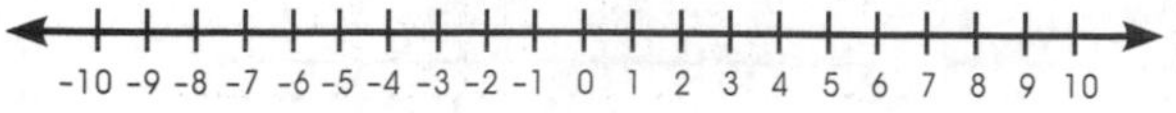

7. ${^-6} \leq {^-2z} \leq 4$

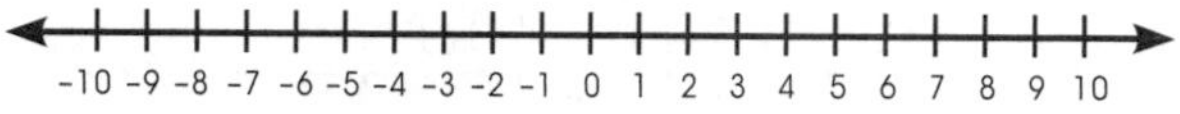

8. $9 \leq 2a + 5 < 15$

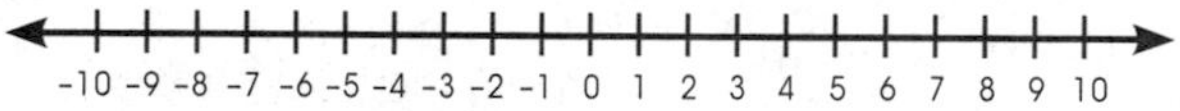

9. $3 < 2x + 1 < 7$

10. ${^-8} < 2x + 4 \leq -2$

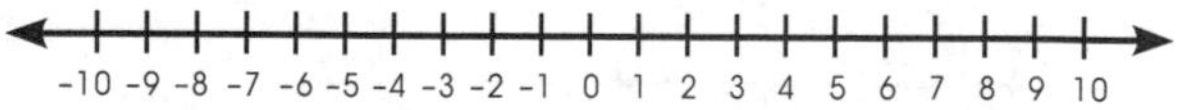

11. ${^-6} \leq 3 - 2(x + 4) \leq 3$

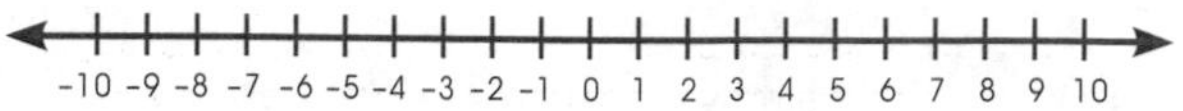

12. $4 - 3x \leq {^-8}$ or $3x - 1 \leq 8$

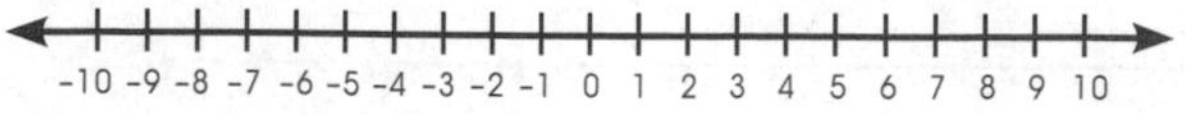

Name________________________________

HSA-REI.D.12

Graphing Linear Equalities

$y < -\frac{1}{2}x + 1$

1.) Graph $y = -\frac{1}{2}x + 1$ as a dotted line.

2.) Choose a point in one half-plane and substitute. Try (0, 3):

$3 < -\frac{1}{2} \cdot 0 + 1 = 3 < 1 =$ False

3.) Shade half-plane that does not contain (0, 3).

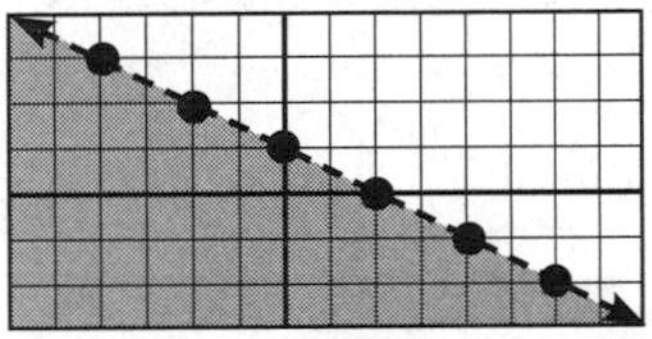

$3x - 4y \leq 12 \Rightarrow y \geq \frac{3}{4}x - 3$

1.) Graph $y \geq \frac{3}{4}$ x − 3 as a solid line.

2.) Choose a point in one half-plane and substitute. Try (0, 0):

$0 \geq \frac{3}{4} \cdot -3 \Rightarrow 0 \geq -3 \Rightarrow$ True

3.) Shade the half-plane that contains (0, 0).

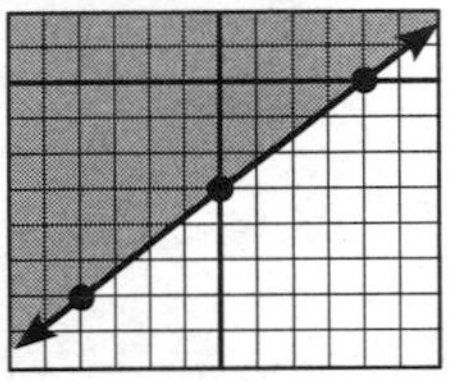

1. $y > x + 1$

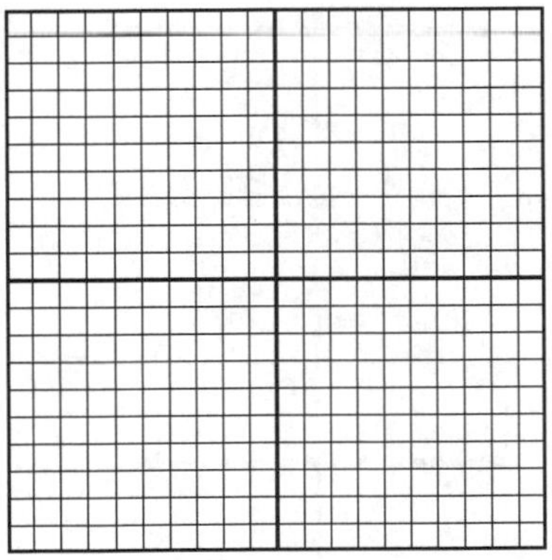

2. $3x - y \leq 6$

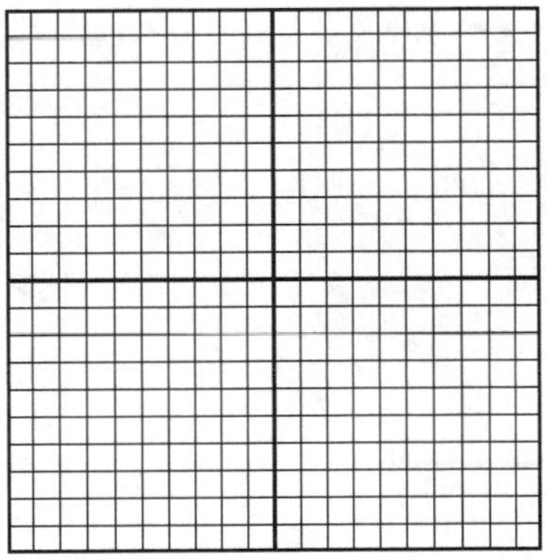

3. $y + 5 \leq 0$

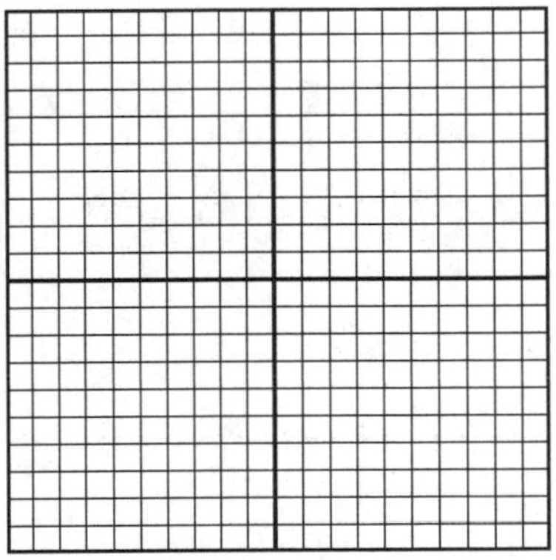

4. $y \geq 2x - 3$

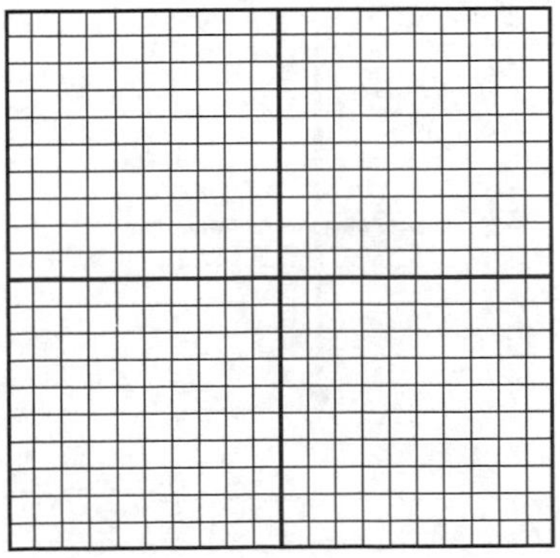

5. $x + y < 3$

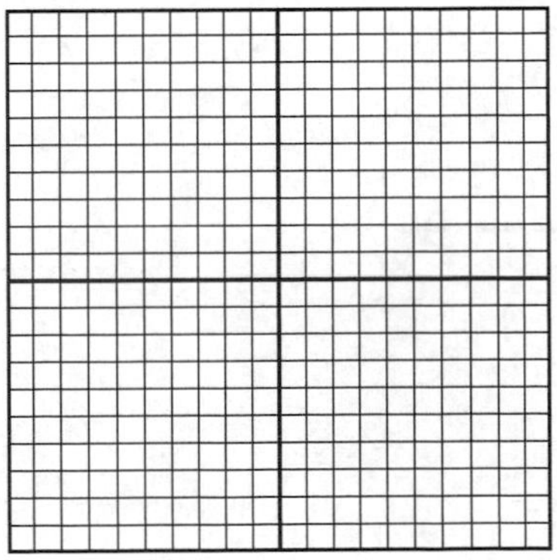

6. $2x + y > {}^{-}8$

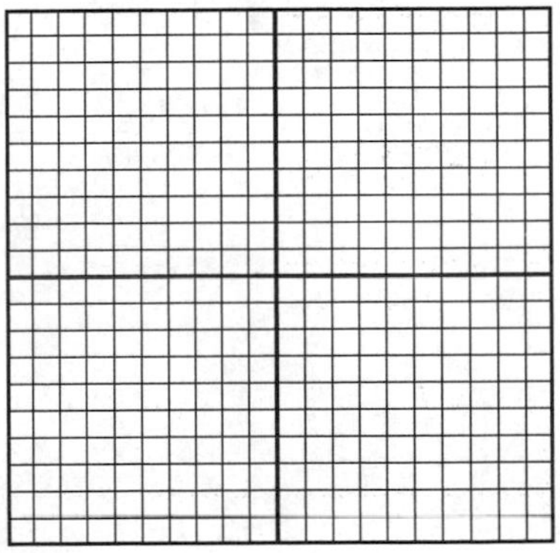

Name________________________________ HSA-REI.D.12

More Graphing Linear Inequalities

Graph each inequality, recalling that ≤ and ≥ require a solid line which includes the values on the line, while < and > require a dashed line, excluding the values on the line.

Example 1: $2x + 6y \geq 12$

x	y
0	2
6	0

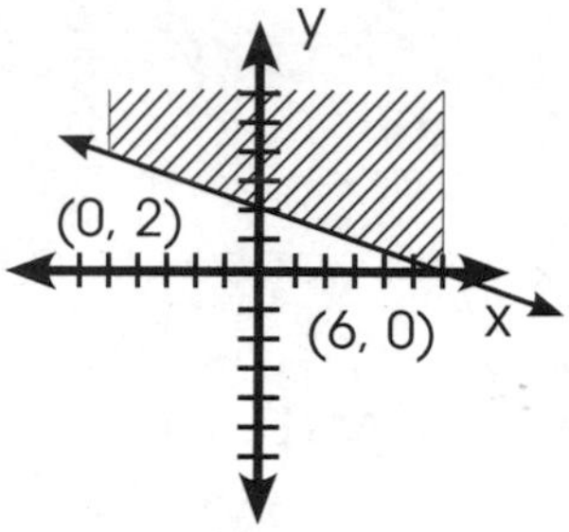

Example 2: $y > {}^{-}2x$

$m = \frac{^{-}2}{1}$ $\frac{\text{Down 2}}{\text{Right 1}}$

y – int (0, 0)

Note: Dashed line

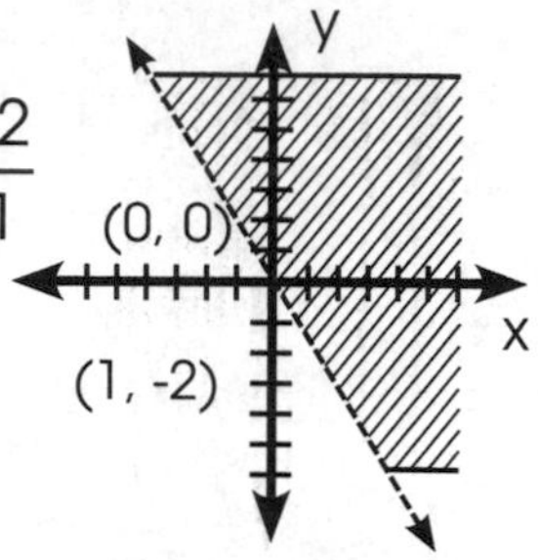

1. $y > x + 1$

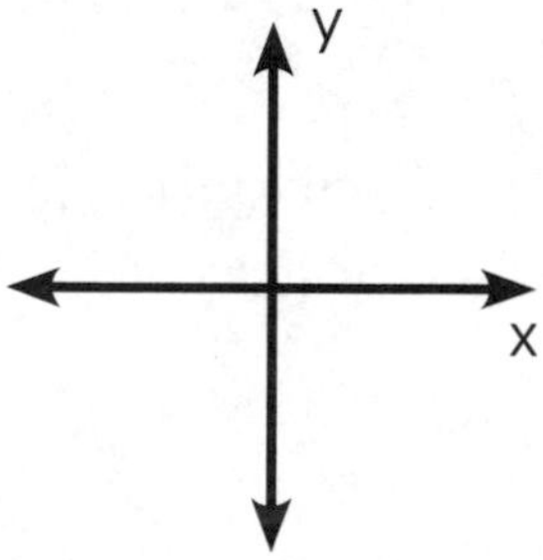

2. $3x - y \leq 6$

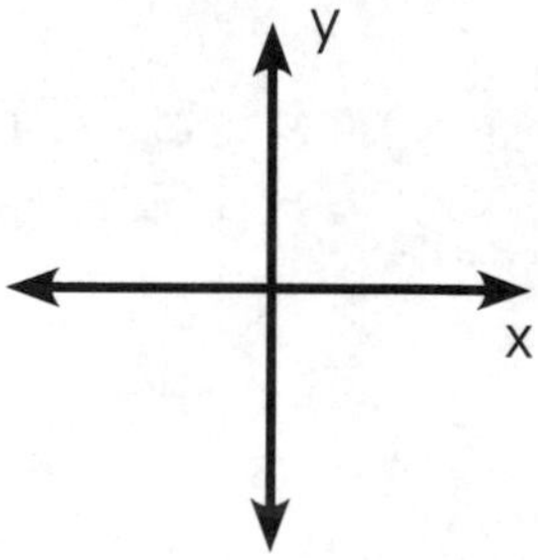

3. $y + 5 \leq 0$

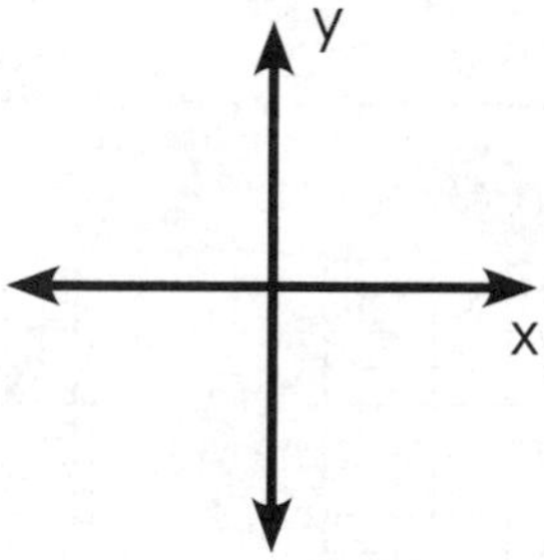

4. $y \geq 2x - 3$

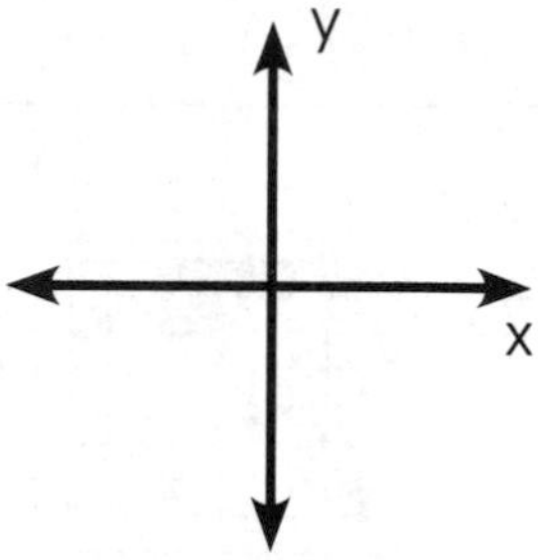

5. $x + y < 3$

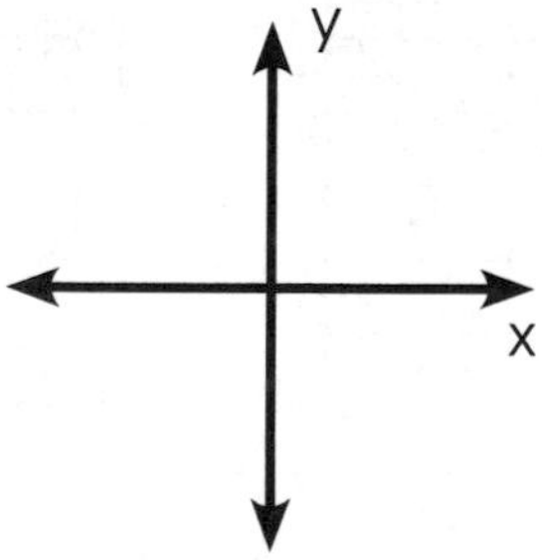

6. $2x + y > {}^{-}8$

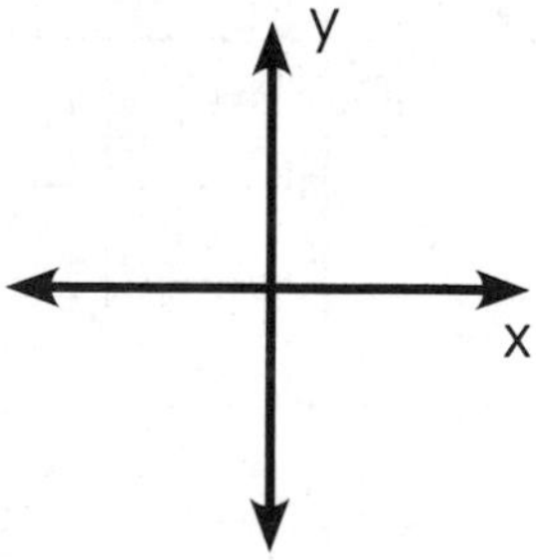

7. $y < -\frac{1}{2}x + 1$

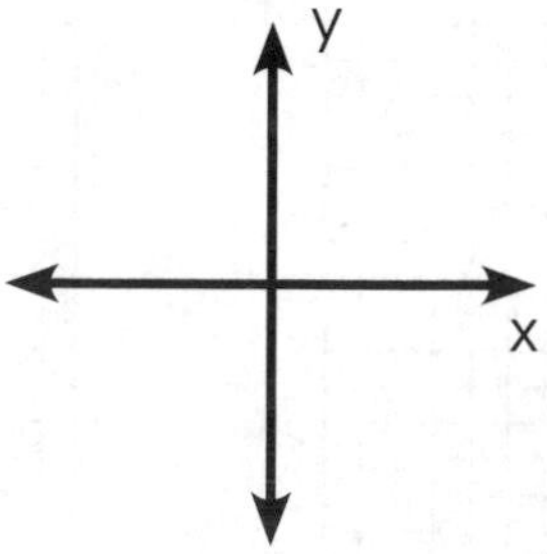

8. $3x - 4y \leq 12$

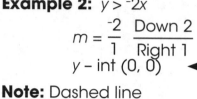

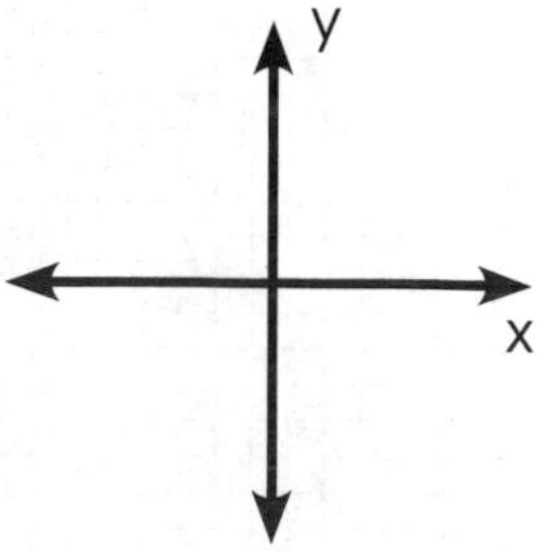

Name________________________ 8.EE.C.8a, HSA-REI.D.12

Graphing Systems of Linear Inequalities

$x + 7 > 2$
$x - 2y \leq {}^{-}7$

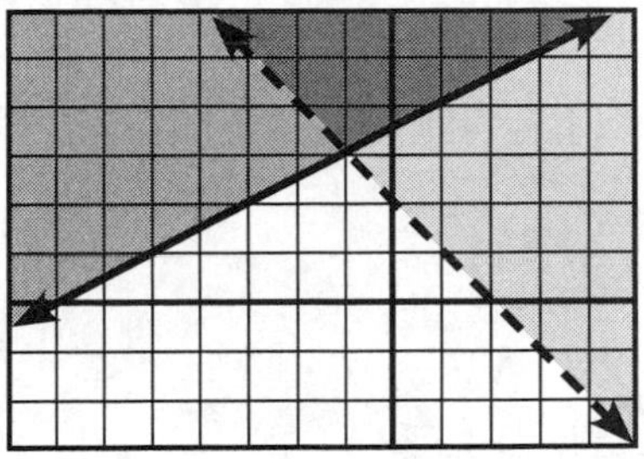

1. $y \geq x - 1$
   $y \leq {}^{-}2x + 1$

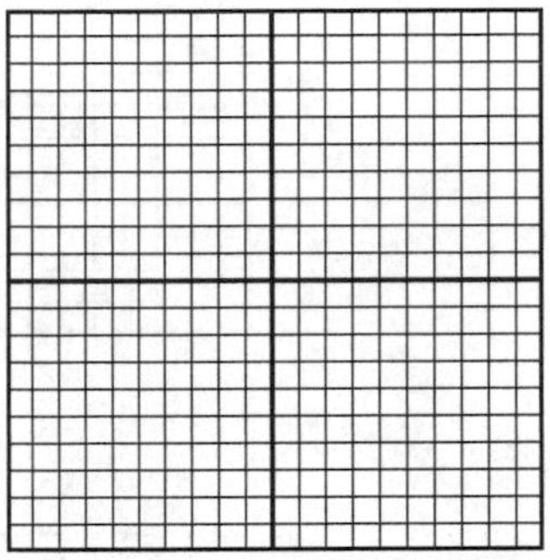

2. $2x - y > 3$
   $x + y > 3$

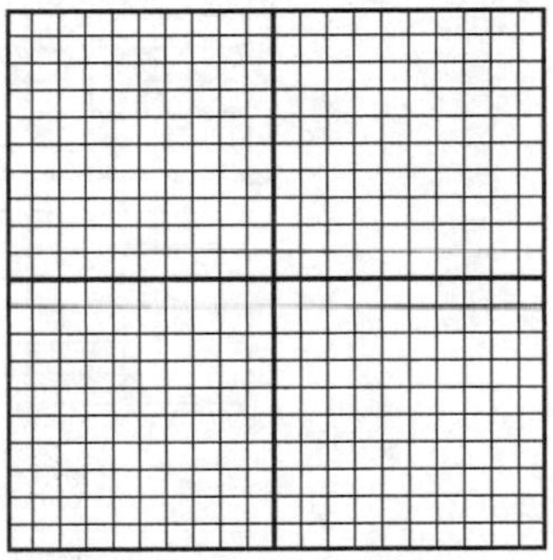

3. $y \geq x$
   $y \geq -x$

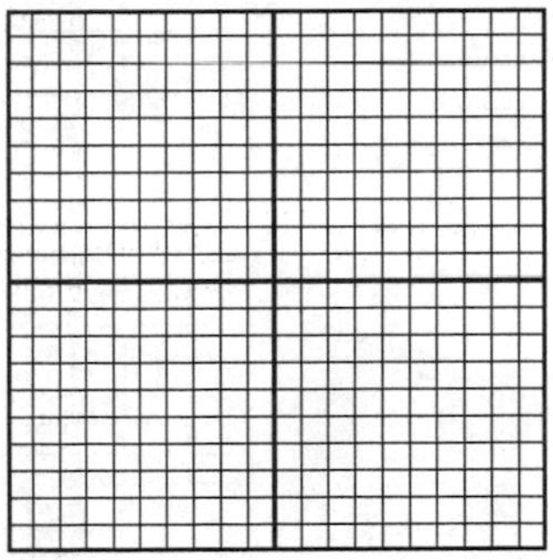

4. $y \leq x + 2$
   $y > 2 - x$

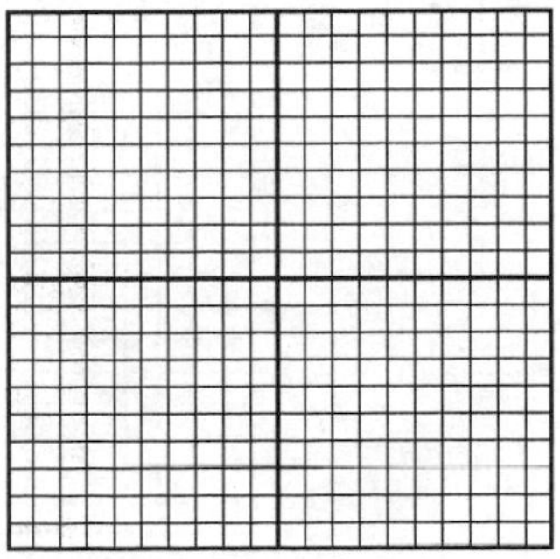

5. $x + y \geq 0$
   $x - y > 0$

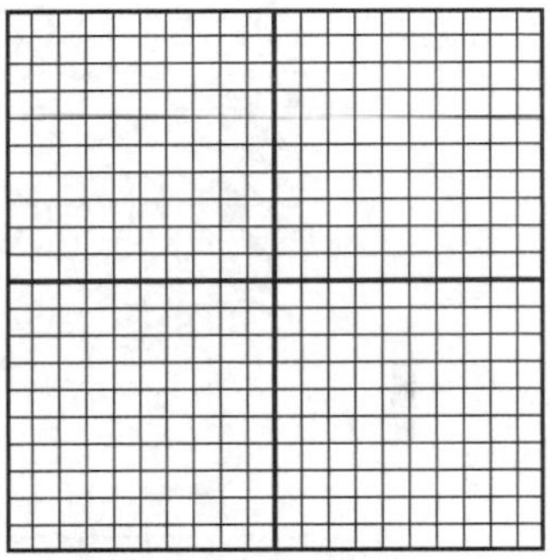

6. $y > \frac{1}{2}x + 3$
   $y > 3$

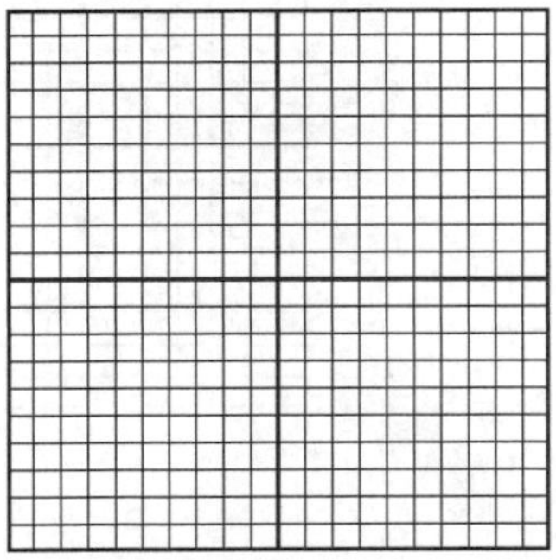

7. $3y - x \geq 3$
   $x \leq {}^{-}2$

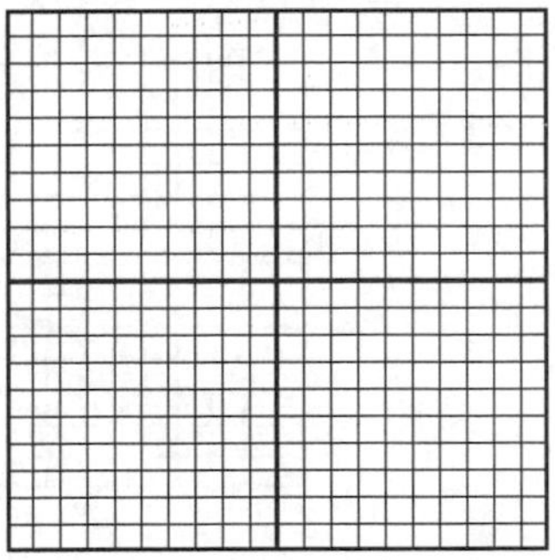

8. $y < 3$
   $x - y < 0$

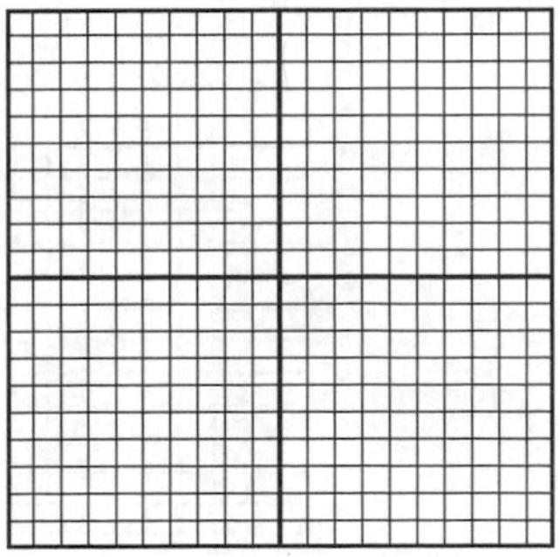

Name____

8.EE.C.8a, HSA-REI.D.12

More Graphing Systems of Linear Inequalities

Graph the systems of linear inequalities.

Example: $\begin{cases} y \geq x - 1 \\ y \leq {}^{-}2x + 1 \end{cases}$

Note: < or > = dashed lines

≤ or ≥ = solid lines

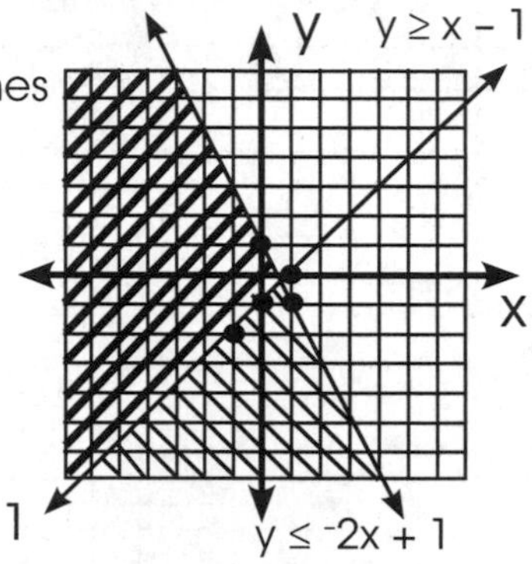

1. $\begin{cases} x - 2y < {}^{-}8 \\ 3x + y > 4 \end{cases}$

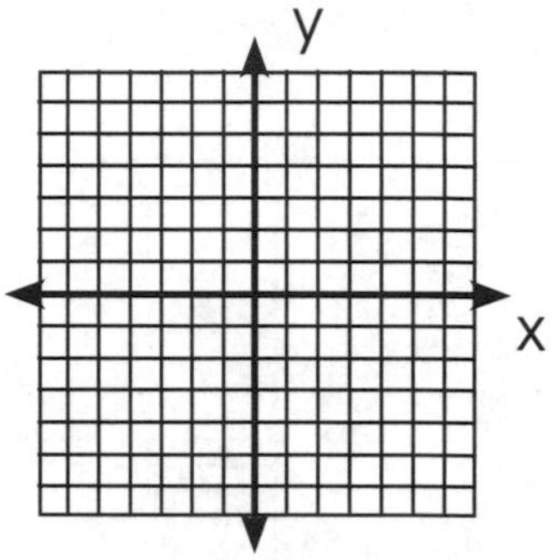

2. $\begin{cases} 3x - 5y > 11 \\ 4x - 3y \geq 5 \end{cases}$

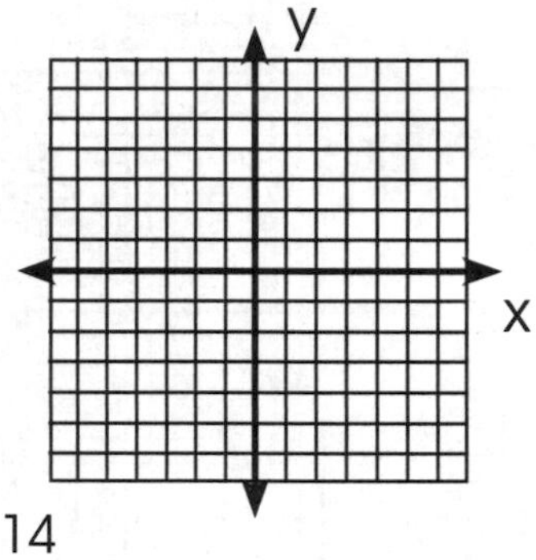

3. $\begin{cases} y > \frac{1}{2}x + 3 \\ y > 3 \end{cases}$

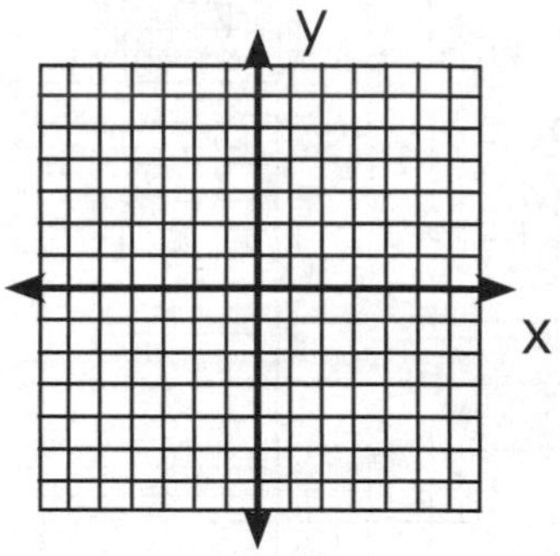

4. $\begin{cases} 2x + 3y \leq 14 \\ 3x - 2y < {}^{-}5 \end{cases}$

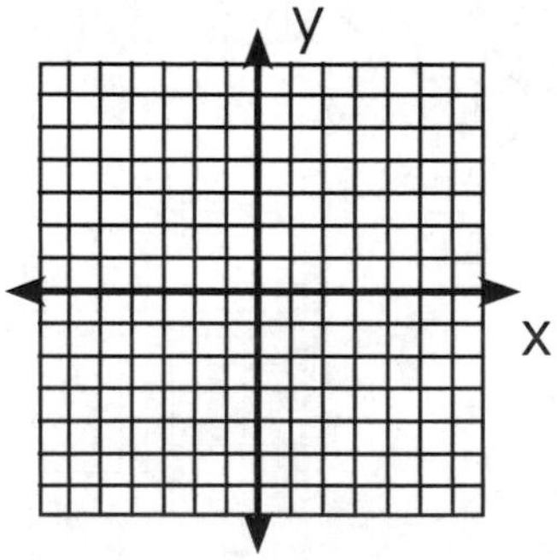

5. $\begin{cases} 5x + 2y > {}^{-}8 \\ 2x - 5y \leq {}^{-}9 \end{cases}$

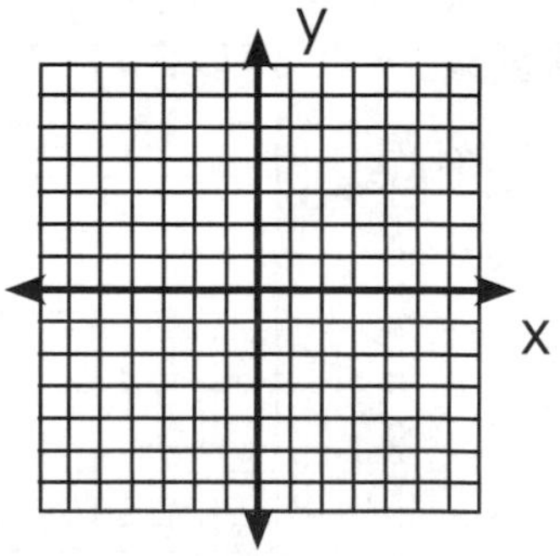

6. $\begin{cases} x \geq 6 \\ y \leq 2x - 6 \end{cases}$

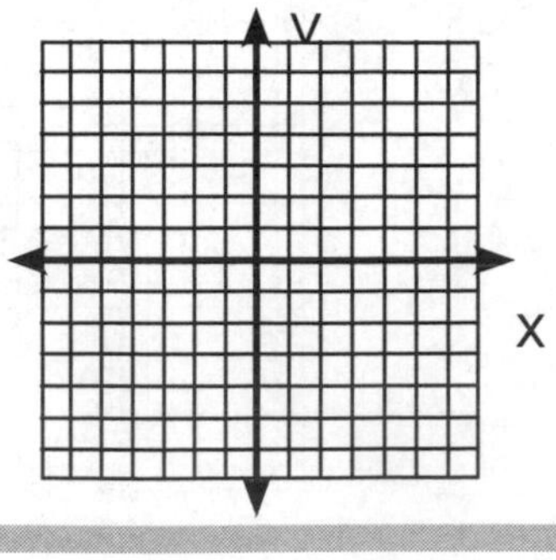

7. $\begin{cases} y > {}^{-}2x + 4 \\ y \leq 6x - 3 \end{cases}$

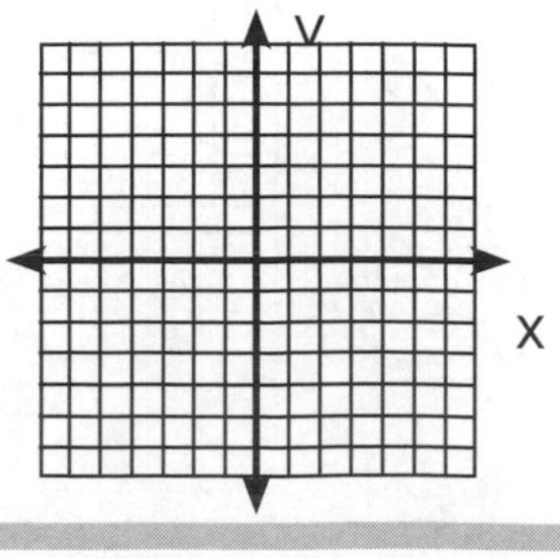

Name________________________________ (8.EE.C.8b, HSA-REI.C.6)

Systems of Equations: Graphic Method

Solve by graphing.

$x - y = {}^{-}2 \Longrightarrow y = x + 2$

$2x + y = 5 \Longrightarrow y = {}^{-}2x + 5$

Solution (1, 3)

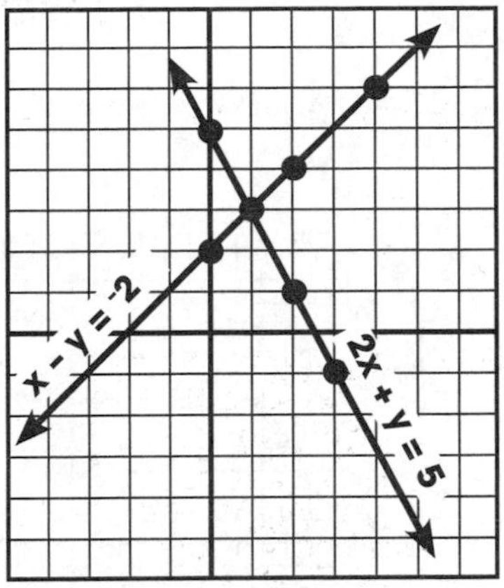

1. $x - y = 6$
   $2x + y = 0$

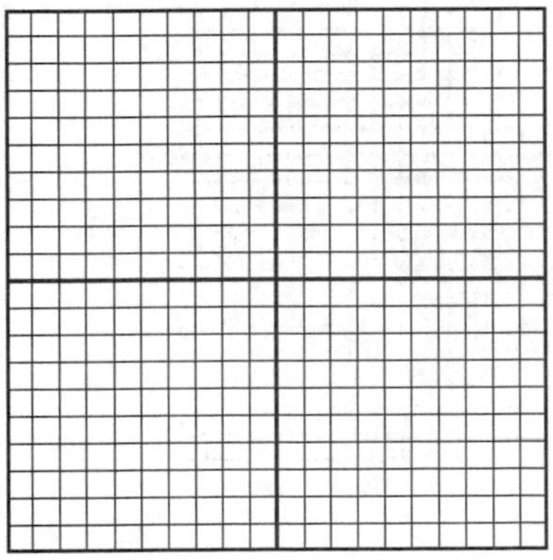

2. $2x - 2y = {}^{-}4$
   $y = 2$

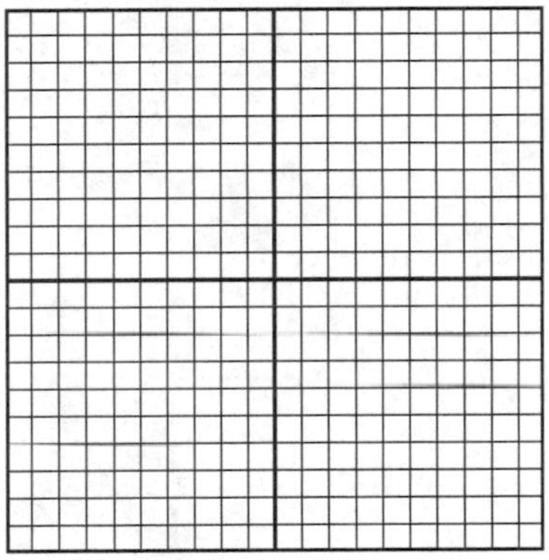

3. $2x - y = 1$
   $3x + y = {}^{-}6$

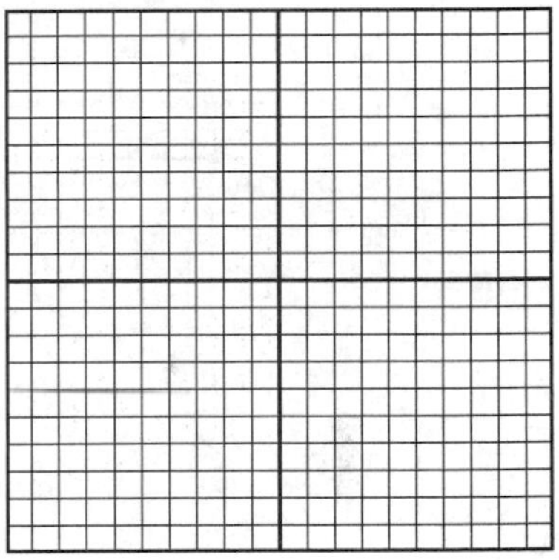

4. $x + 2y = 4$
   $2x - y = 8$

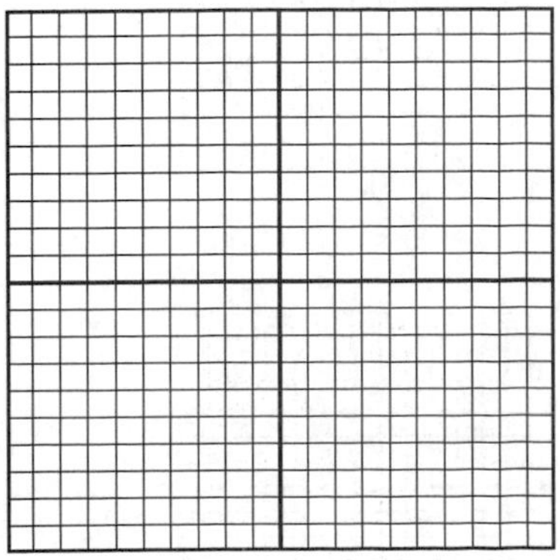

5. $2x - y = 5$
   $x - y = 1$

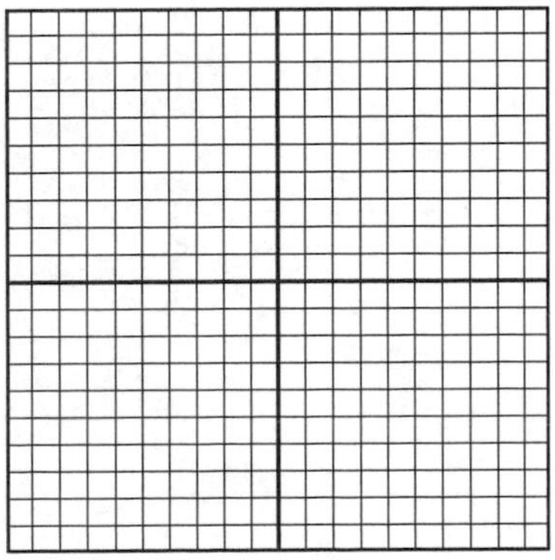

6. $x = 3$
   $y = {}^{-}2$

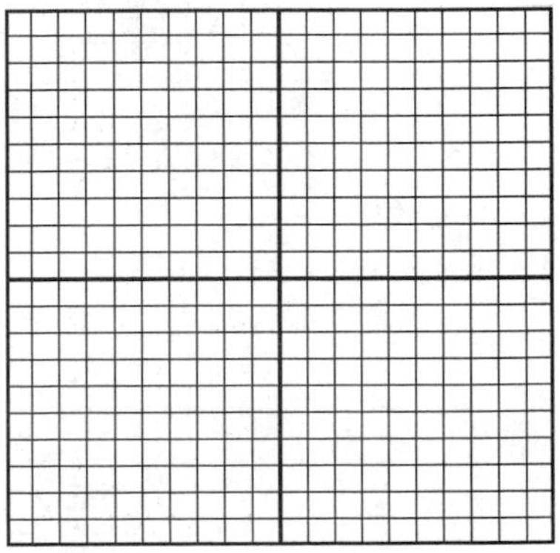

Name________________________________

8.EE.C.8b, HSA-REI.C.6

Solving Systems by Graphing

Estimate the solution of the systems by graphing.

Example: $\begin{cases} x - y = 6 \\ 2x + y = 0 \end{cases}$

x	y
0	$^-6$
6	0

x	y
0	0
1	$^-2$

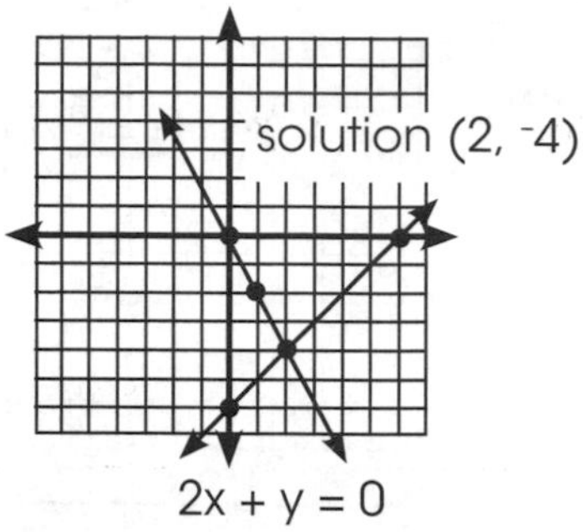

1. $\begin{cases} y = 6 \\ 2x - 2y = {}^-4 \end{cases}$

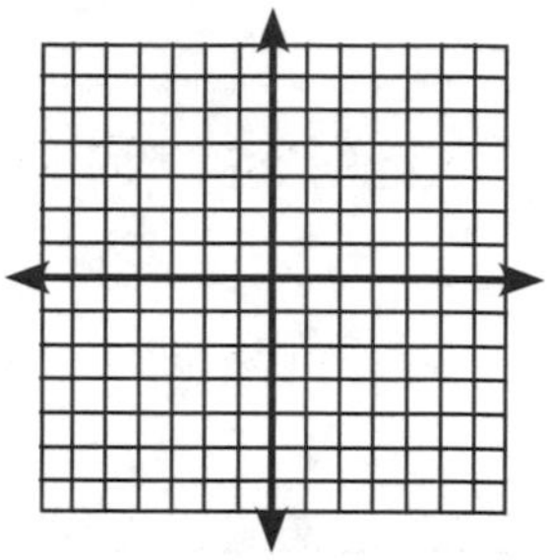

2. $\begin{cases} x = 4 \\ y = 2x \end{cases}$

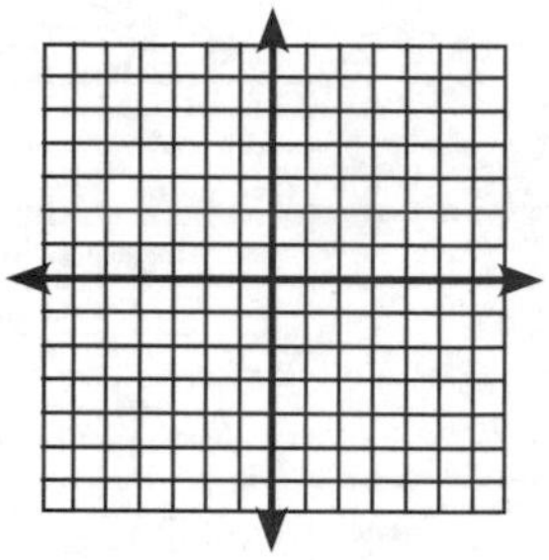

3. $\begin{cases} 2x - y = 1 \\ 3x + y = {}^-6 \end{cases}$

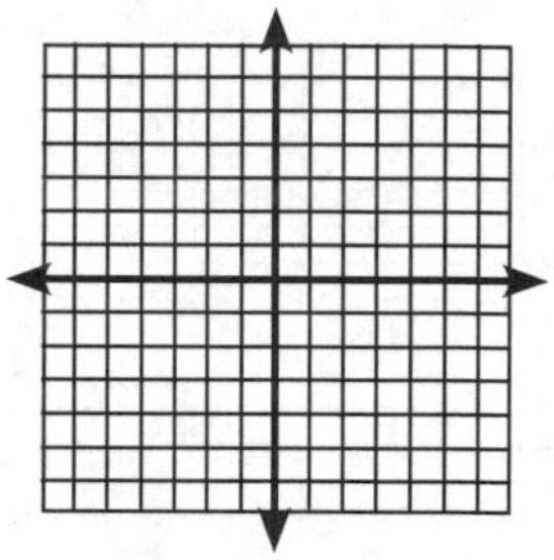

4. $\begin{cases} 2x - y = 5 \\ x - y = 1 \end{cases}$

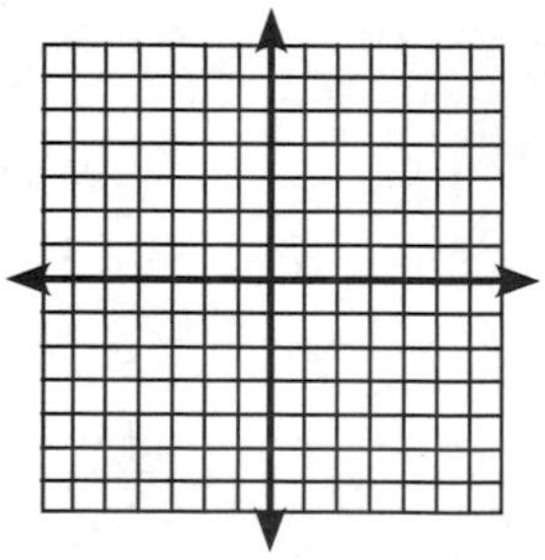

5. $\begin{cases} x = 2 \\ y = {}^-6 \end{cases}$

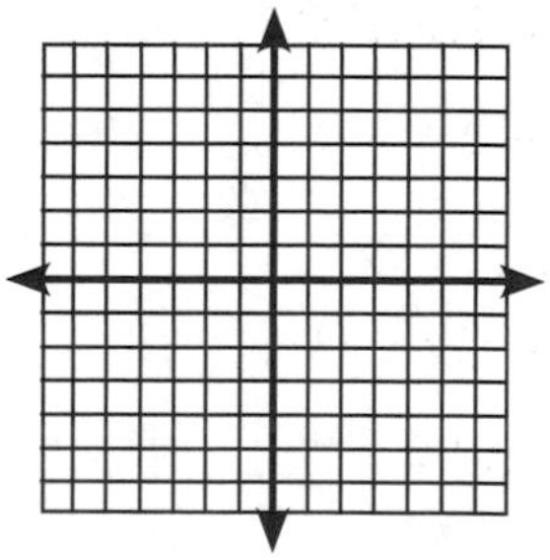

6. $\begin{cases} 2x + y = {}^-6 \\ 3x + y = {}^-10 \end{cases}$

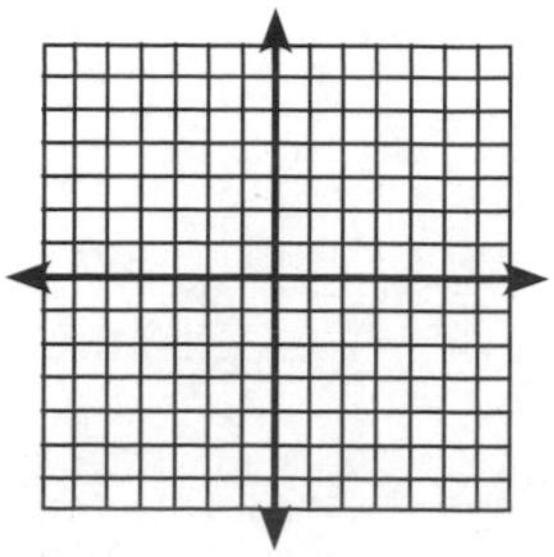

7. $\begin{cases} 7y + 15 = 3x \\ 15 = 3x + 2y \end{cases}$

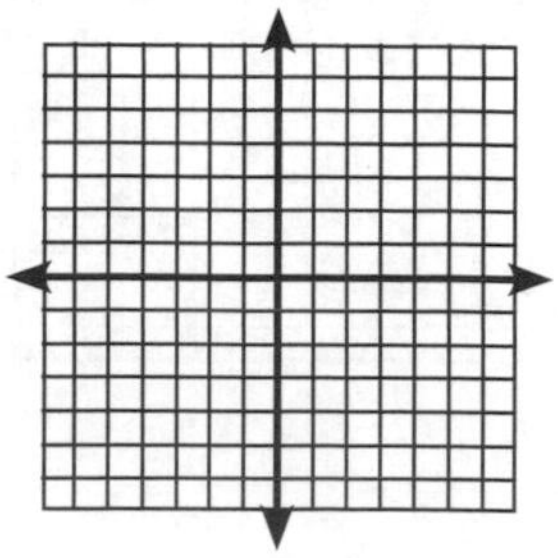

Name________________________________ 8.EE.C.8b, HSA-REI.C.6

Systems of Equations: Elimination Method

$x + y = 6$
$x - y = 4$

$$\begin{array}{r} x + y = 6 \\ + x - y = 4 \\ \hline 2x \quad = 10 \\ x = 5 \end{array}$$

$x + y = 6 \Rightarrow 5 + y = 6$
$y = 1$

Solution (5, 1)

$3y = {}^-7x + 7$
$2y = 7x - 7$

$$\begin{array}{r} 7x + 3y = 7 \\ -(7x - 2y = 7) \\ \hline 5y = 0 \\ y = 0 \end{array}$$

$2y = 7x - 7 \Rightarrow 0 = 7x - 7$
$7 = 7x$
$1 = x$

Solution (1, 0)

1. $2x + y = {}^-6$
   $3x + y = {}^-10$

2. $8x - y = 20$
   ${}^-5x + y = {}^-8$

3. $2x + y = 0$
   $2x - 3y = {}^-8$

4. $5x + 3y = 10$
   $2x - 3y = 4$

5. $9x - 3y = 9$
   $x + 3y = 11$

6. $x + 3y = 9$
   $x - 2y = {}^-6$

7. $2x + y = 4$
   $2x + 2y = 2$

8. $7y + 15 = 3x$
   $15 = 3x + 2y$

9. $25x = 91 - 16y$
   $16y = 64 - 16x$

10. $4x - 2y = {}^-2$
    $4x + 3y = {}^-12$

11. $2x + y = {}^-7$
    $y = 3x + 3$

12. $3x = {}^-2y + 10$
    $x = 2y + 6$

13. $x + 4y = 2$
    $x - 2y = 8$

14. $x + 5y + 11 = 0$
    $3x - 5y - 7 = 0$

Name________________________________ 8.EE.C.8b, HSA-REI.C.6

Solving Systems by Elimination

Create a quick graph and solve the system by elimination.

Hint: Rewrite one or both of the equations so that adding "eliminates" a variable.

Example: $\begin{cases} 3x - 4y = {}^-15 \\ 5x + y = {}^-2 \end{cases}$

x	y
0	$\frac{15}{4} \approx 3.75$
$^-5$	0

x	y
0	$^-2$
$-\frac{2}{5}$	0

$3x - 4y = {}^-15$
$4(5x + y = {}^-2)$

$3x - 4y = {}^-15$
$\underline{20x + 4y = {}^-8}$
$23x = {}^-23$
$x = {}^-1$

$5x + y = {}^-2$
$y = {}^-5\,({}^-1) - 2$
$y = 3$

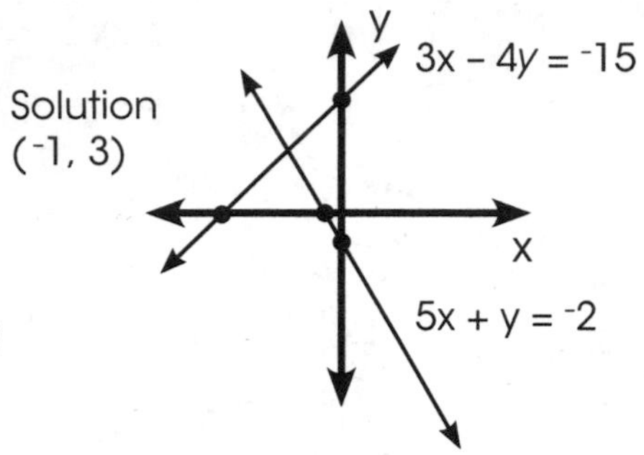

1. $\begin{cases} 2x + y = 0 \\ {}^-2x + 3y = 8 \end{cases}$

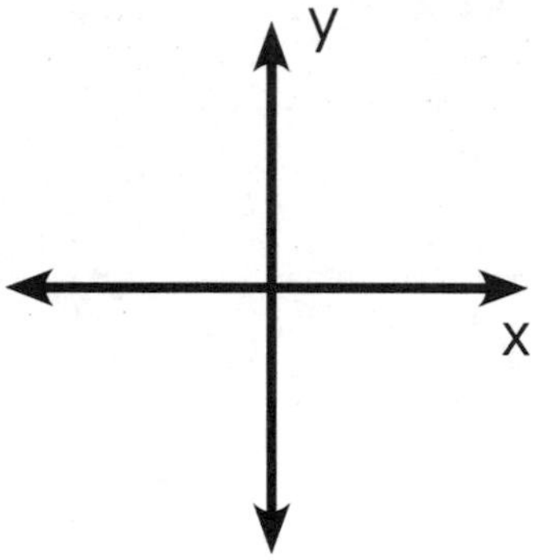

2. $\begin{cases} 5x - 6y = 16 \\ 5x + y = 2 \end{cases}$

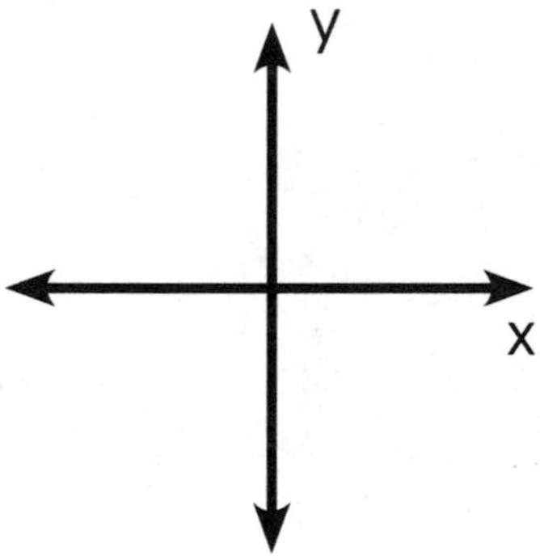

3. $\begin{cases} 4x - 2y = {}^-2 \\ 4x + 3y = {}^-12 \end{cases}$

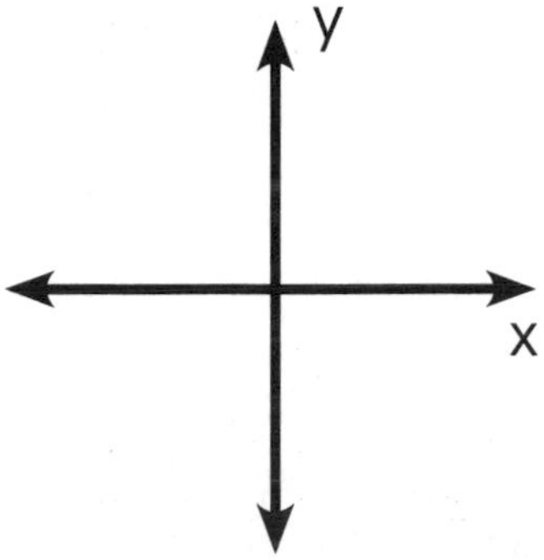

4. $\begin{cases} x - 2y = {}^-8 \\ 3x + y = 4 \end{cases}$

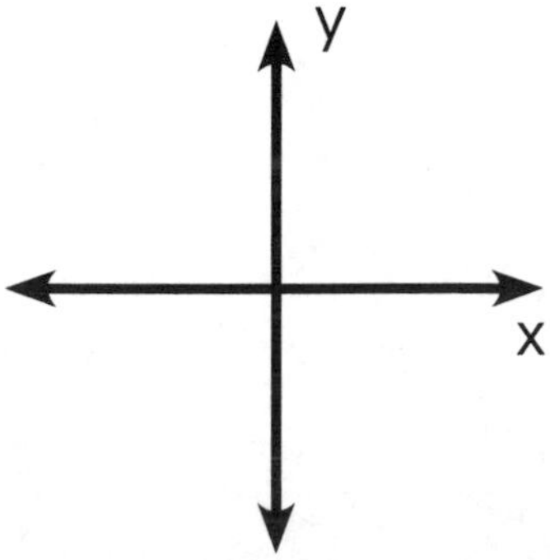

Name________________________________ HSA-REI.C.5

Systems of Equations: Substitution Method

$x - 5 = 10$

${}^{-}2x + y = 7 \Rightarrow y = 2x + 7$

Solution $({}^{-}5, {}^{-}3)$

$x - 5(2x + 7) = 10$

$x - 10x - 35 = 10$

${}^{-}9x - 35 = 10$

${}^{-}9x = 45$

$x = {}^{-}5$

1. $y = 5 - 4x$
   $3x - 2y = 12$

2. $3x + 2y = 8$
   $x = 3y + 10$

3. $3x - 4y = {}^{-}15$
   $5x + y = {}^{-}2$

4. $x + y = 2$
   $3x + 2y = 5$

5. $x = 3 - 3y$
   $4y = x + 11$

6. $x - y = {}^{-}15$
   $x + y = {}^{-}5$

7. $2x + y = {}^{-}6$
   $3x + y = {}^{-}10$

8. $y = -x + 6$
   $x - 2y = {}^{-}6$

9. $2y - x = 6$
   $3y - x = 4$

10. $5x - 6y = 16$
    $5x + y = 2$

11. $y = 3x$
    $x + y = 8$

12. $x - 3y = {}^{-}5$
    $2x + y = 11$

13. $-x + y = 5$
    $y = {}^{-}3x + 1$

14. $2x = 3y$
    $x = 3y - 3$

Name______________________________ HSA-REI.C.5

Solving Systems by Substitution

Create a quick graph and solve the system by substitution.

Example: $\begin{cases} y = -x + 6 \\ x - 2y = {}^{-}6 \end{cases}$

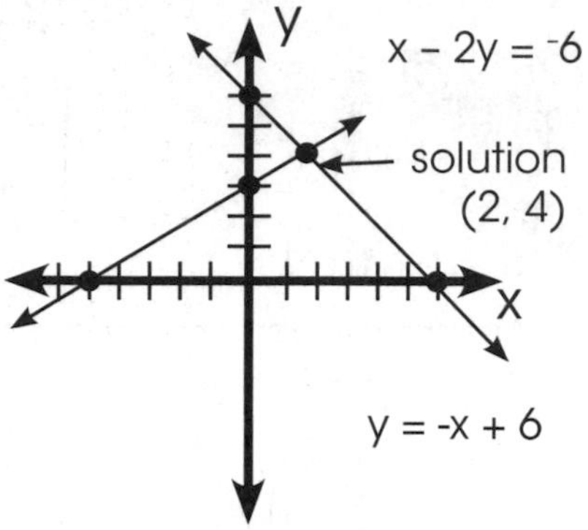

x	y
0	3
$^{-}6$	0

graph: y – int (0, 6)

$m = \frac{^{-}1}{1}$ $\frac{\text{down one}}{\text{right one}}$

solve:

$x - 2(-x + 6) = {}^{-}6$

$x + 2x - 12 = {}^{-}6$ $\quad y = -x + 6$

$3x = 6$ $\quad y = {}^{-}2 + 6$

$x = 2$ $\quad y = 4$

1. $\begin{cases} y = 3x \\ x + y = 8 \end{cases}$

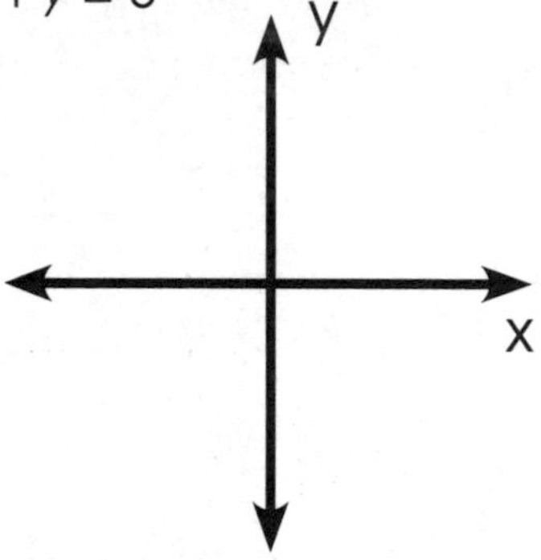

2. $\begin{cases} 2x = 3y \\ x = 3y - 3 \end{cases}$

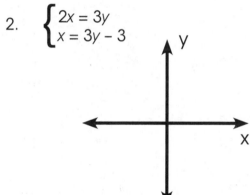

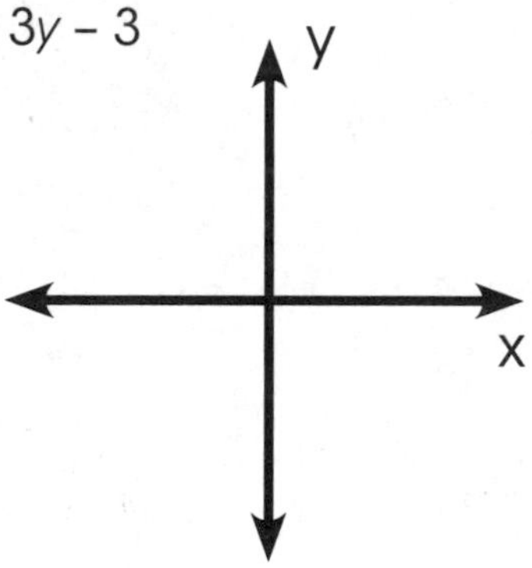

3. $\begin{cases} y = x - 2 \\ x + y = 0 \end{cases}$

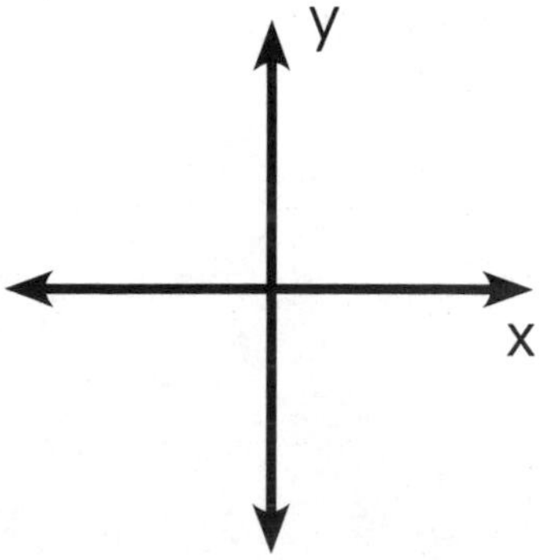

4. $\begin{cases} \frac{3}{2}x - 2y = 20 \\ y = \frac{^{-}5}{4}x \end{cases}$

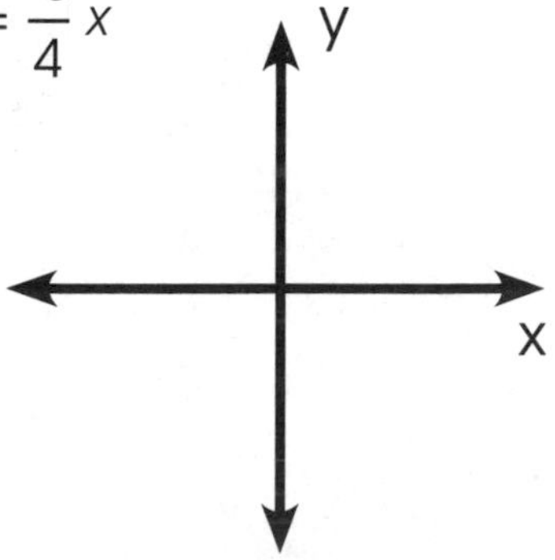

Name________________________________ (8.EE.C.8b, HSA-REI.C.5)

Review: Solving Linear Systems

Create a quick graph and solve using elimination or substitution.

1. $\begin{cases} 2x - 4y = 18 \\ 3x - y = 22 \end{cases}$

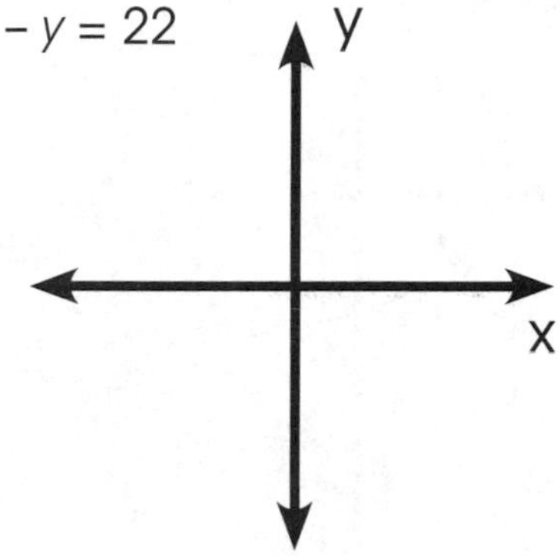

2. $\begin{cases} 3x + 10y = 16 \\ x = {}^{-}6y \end{cases}$

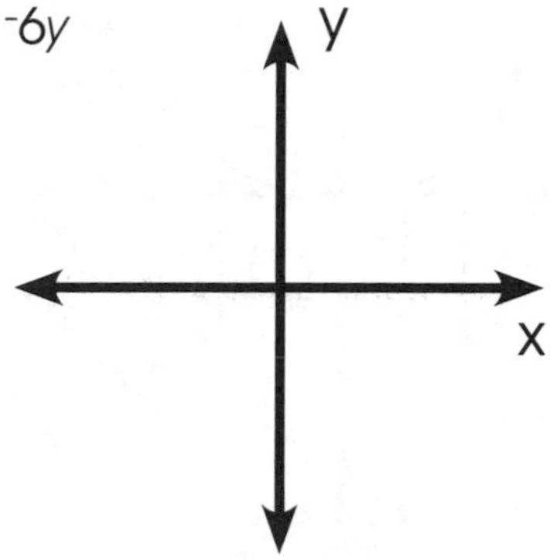

3. $\begin{cases} y = 5x \\ {}^{-}3x + 2y = {}^{-}28 \end{cases}$

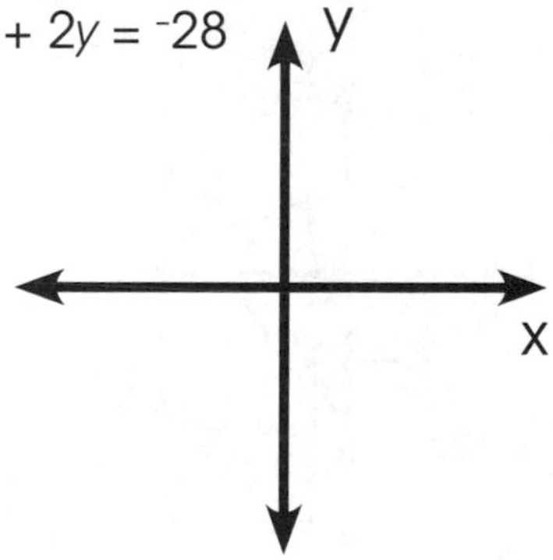

4. $\begin{cases} 4x - 5y = {}^{-}19 \\ 3x + 7y = 18 \end{cases}$

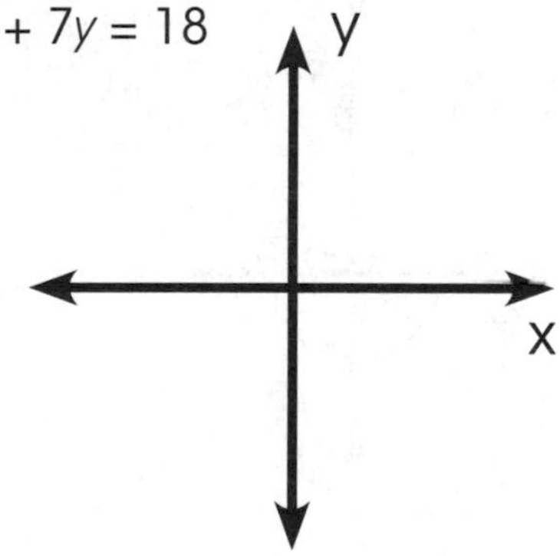

5. $\begin{cases} y = 4 \\ 5x - 6y = 11 \end{cases}$

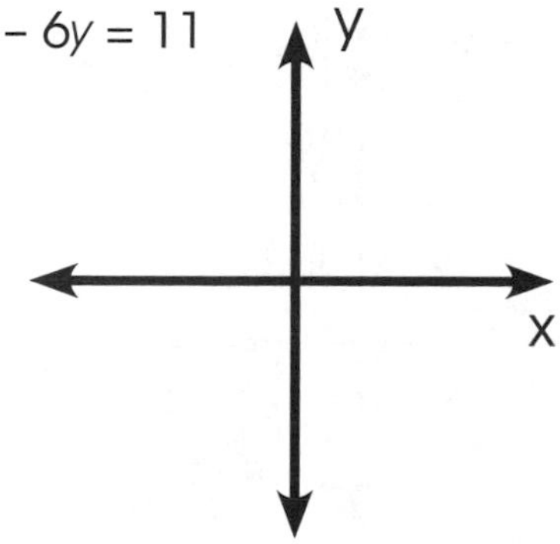

6. $\begin{cases} y = x - 4 \\ 2x - y = {}^{-}2.5 \end{cases}$

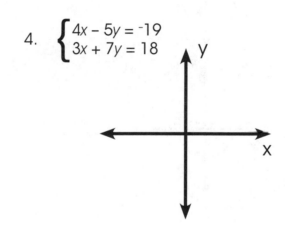

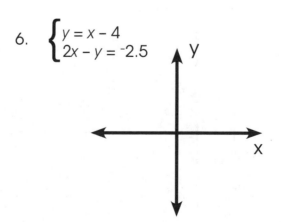

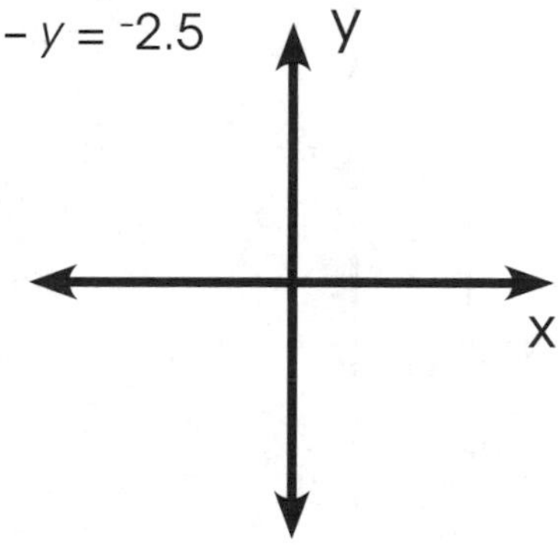

Name________________________________ (8.F.A.1)

Using the Vertical Line Test

Determine whether or not each graph is a function by applying the vertical line test.
Remember: A graph represents a function if every *x* is associated with a unique *y*.

Example: YES

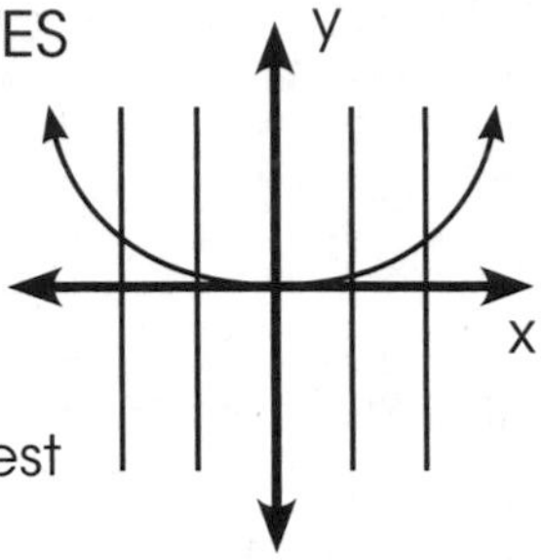

vertical line test

Example: NO

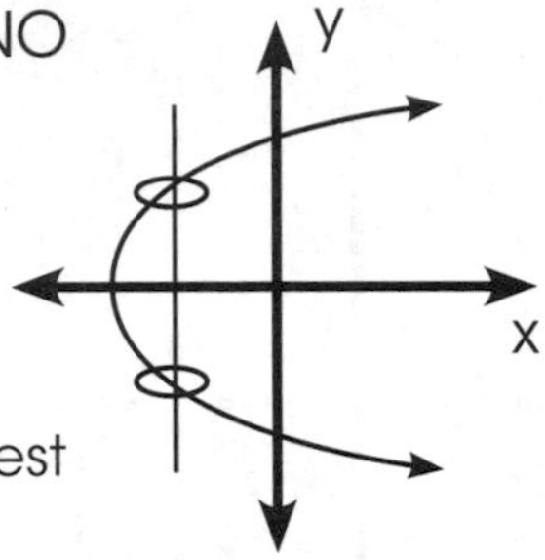

vertical line test

1.

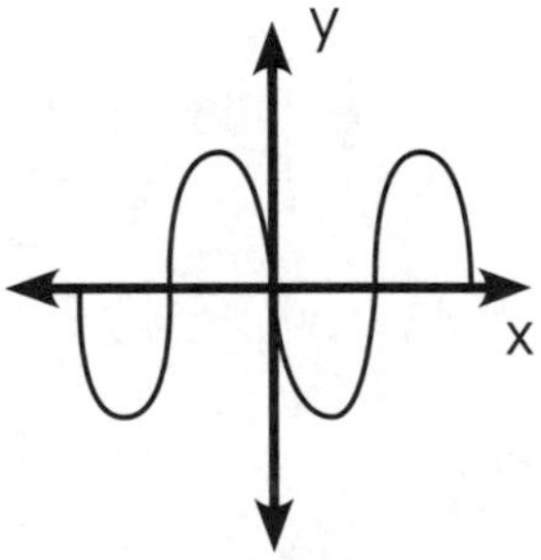

2.

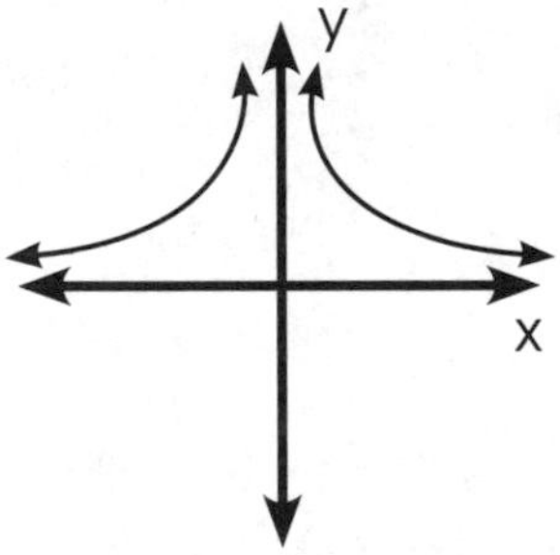

3.

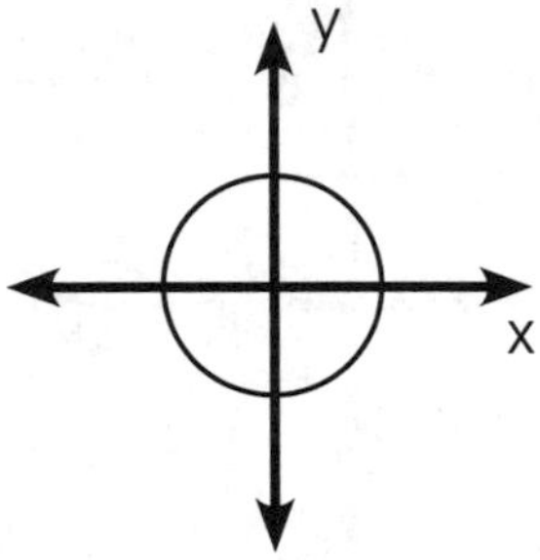

4.

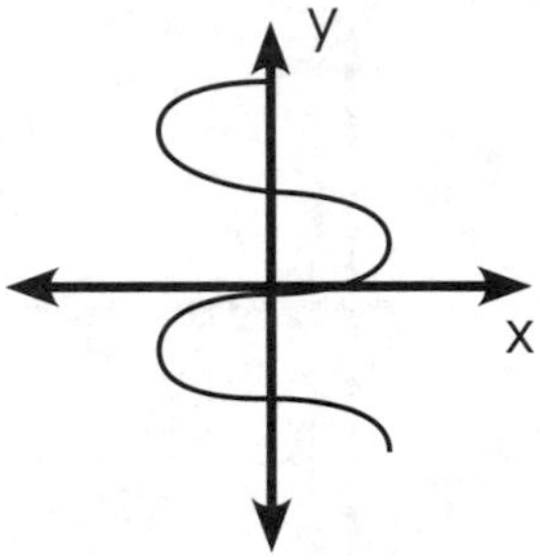

5.

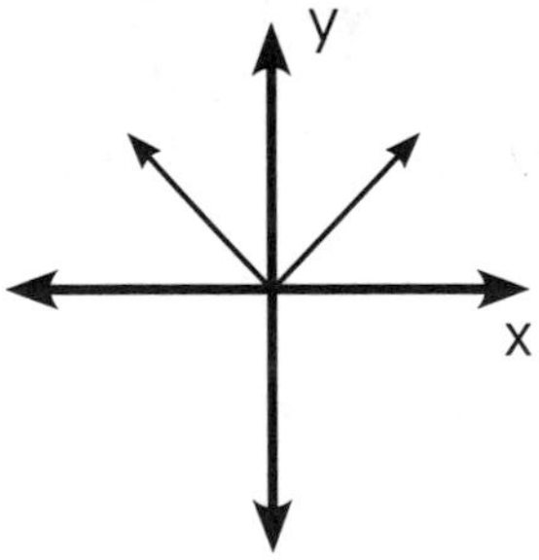

6.

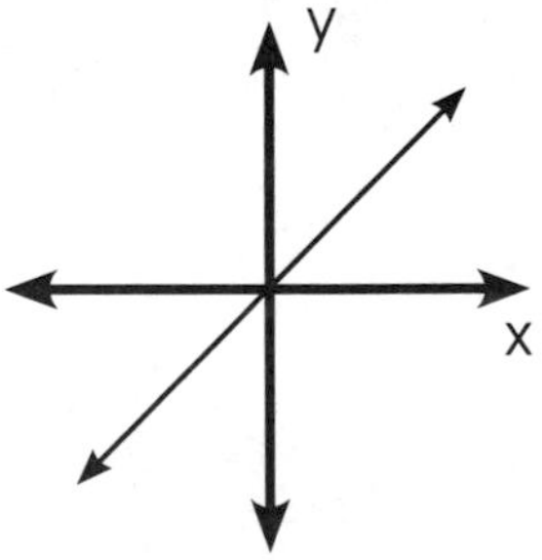

7.

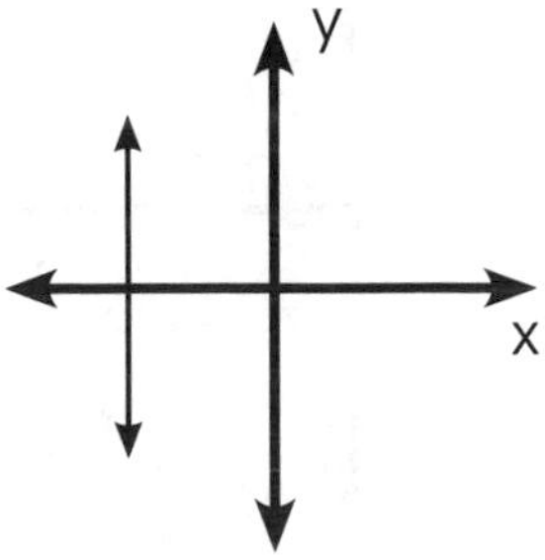

8.

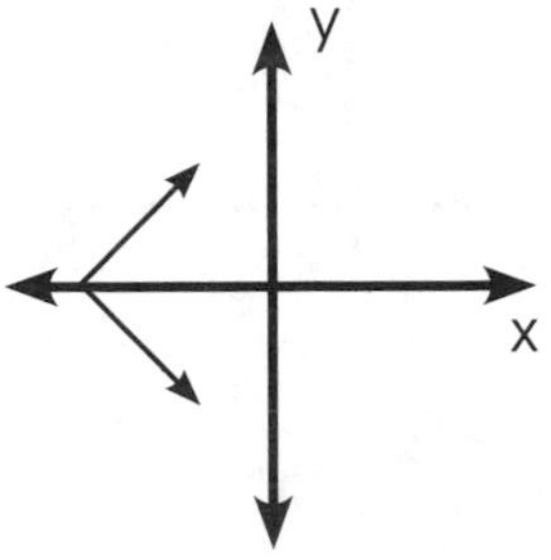

Name________________________________ HSF-IF.A.2

Compositions of Functions

Write the composition of the functions given.

Given: $f(x) = {}^{-}2x + 7$ $g(x) = x^2$ $h(x) = x - 1$

Example: $g(h(x))$

$g(h(x)) = (x - 1)^2$

$= x^2 - x - x + 1$

$g(h(x)) = x^2 - 2x + 1$

1. $f(g(x))$

2. $g(f(x))$

3. $h(g(x))$

4. $f(h(x))$

5. $h(f(x))$

6. $f(g(h(x)))$

7. $h(g(f(x)))$

8. $f(f(x))$

Name________________________________ 8.F.A.2, HSF-IF.A.1, HSA-REI.D.10

Using a Table of Values

Complete each table of values and graph the function.

$y = 2x - 1$ Euler notation: $f(x) = y$ so $f(x) = 2x - 1$

Example: $f(x) = {}^{-}3x + 2$

x	$f(x)$
-1	5
0	2
1	-1

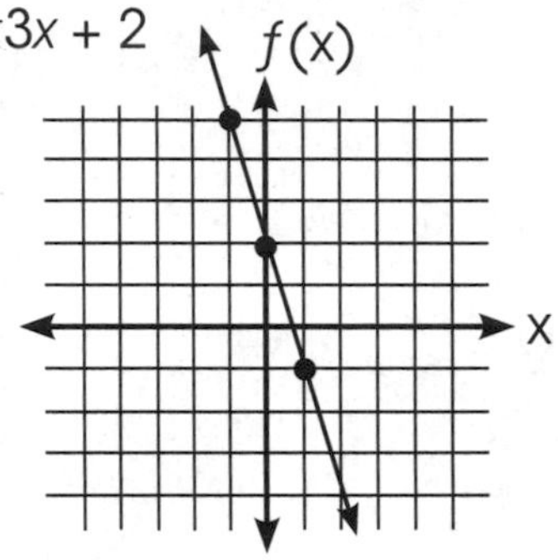

1. $f(x) = -x^2 + 3$

x	$f(x)$

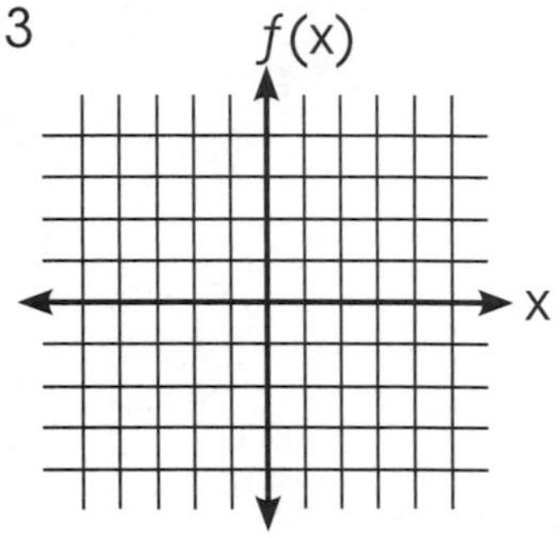

2. $f(x) = \frac{-1}{2}x + 1$

x	$f(x)$

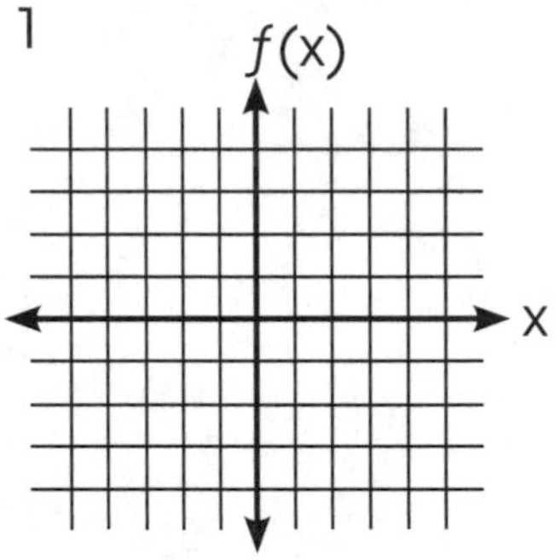

3. $f(x) = x^2 - 2x + 1$

x	$f(x)$

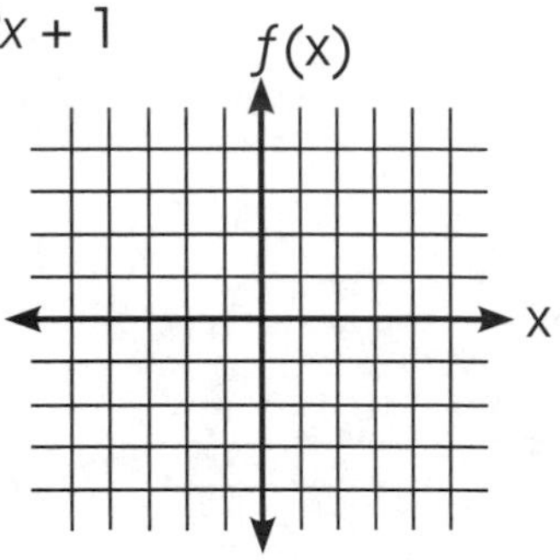

4. $f(x) = 2x^2 - 4$

x	$f(x)$

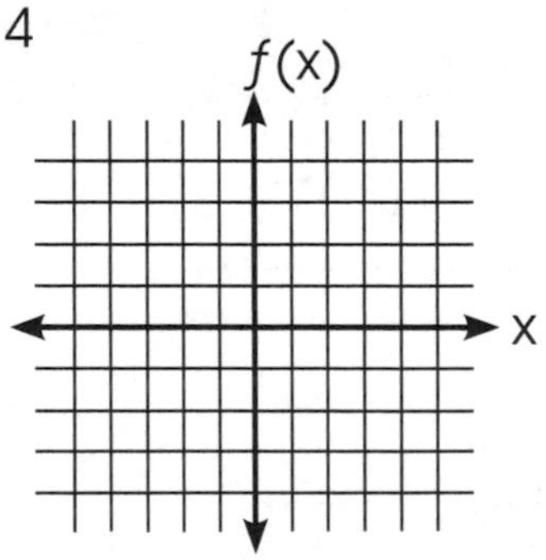

5. $f(x) = 2x - 5$

x	$f(x)$

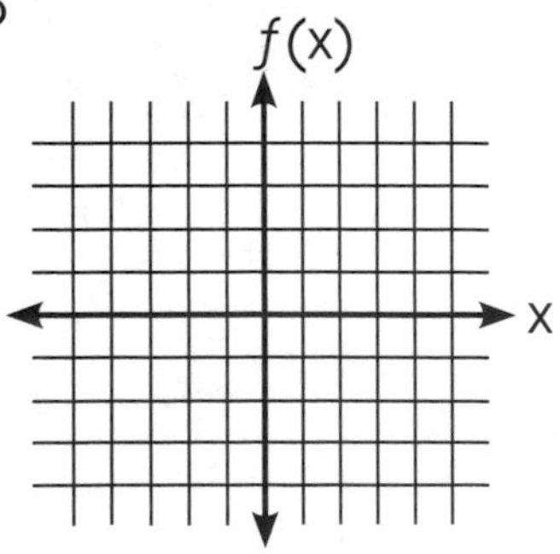

6. $f(x) = |x|$

x	$f(x)$

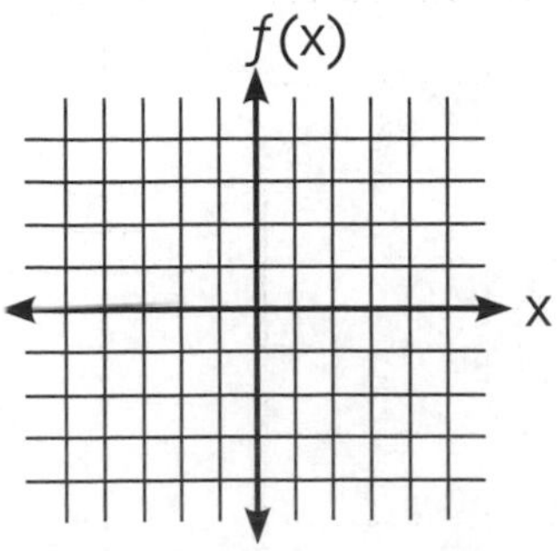

7. $f(x) = x^3$

x	$f(x)$

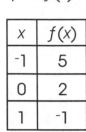

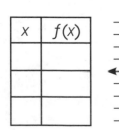

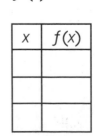

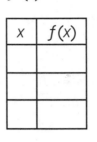

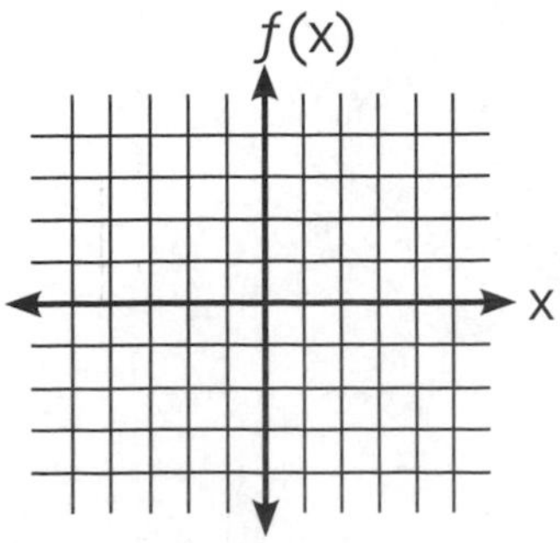

Name____

HSF-BF.B.4c

Graphing the Inverse of a Function

Fill in each table of values, then graph the inverse of each function.

Example:

function

x	$f(x)$
-2	4
-1	1
-1/2	1/4
0	0
1/2	1/4
1	1
2	4

inverse

x	$f(x)$
4	-2
1	-1
1/4	-1/2
0	0
1/4	1/2
1	1
4	2

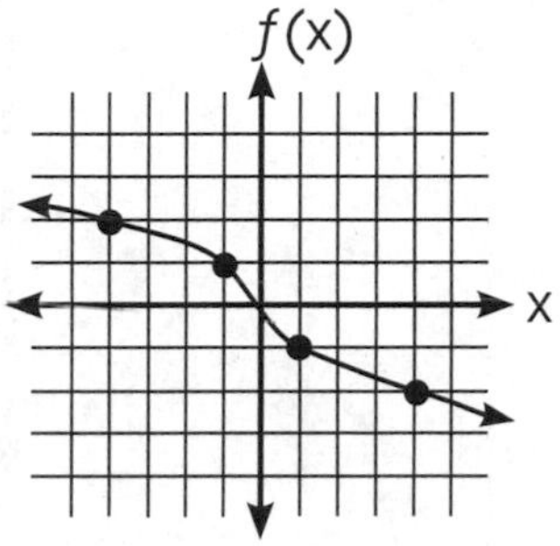

1.

function

x	$f(x)$
-2	2
-1	1
-1/2	1/2
0	0

inverse

x	$f(x)$

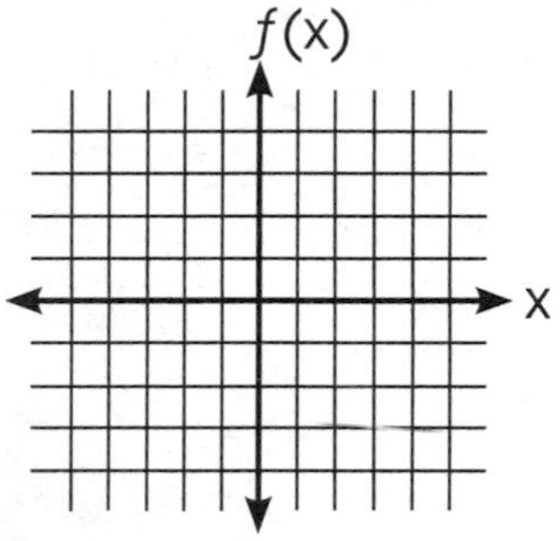

2.

function

x	$f(x)$
-2	-8
-1	-1
-1/2	-1/8
0	0
1/2	1/8
1	1
2	8

inverse

x	$f(x)$

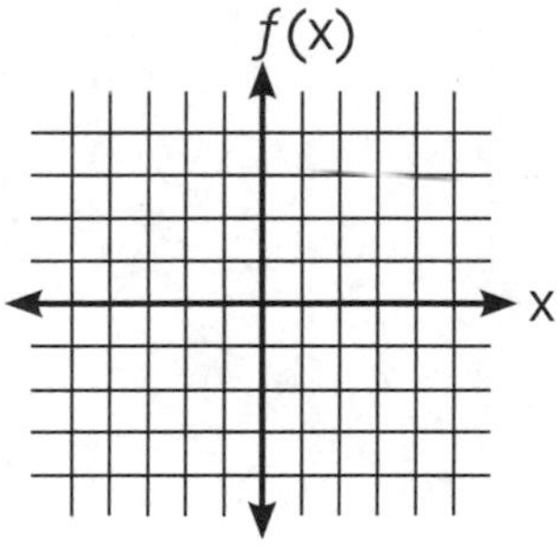

3.

function

x	$f(x)$
-3	-11
-2	-9
-1	-7
0	-5
1	-3
2	-1
3	1

inverse

x	$f(x)$

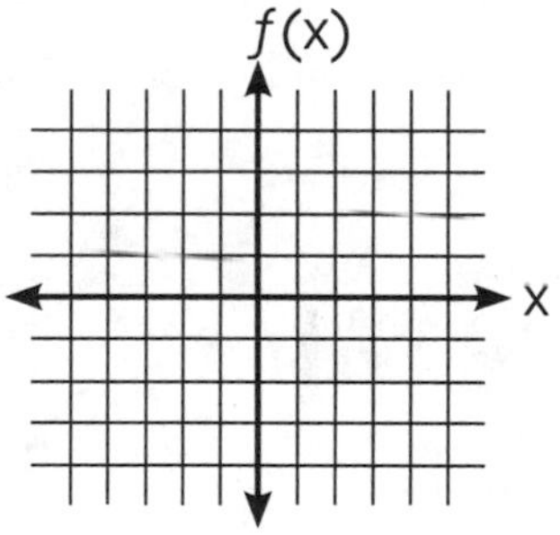

4.

function

x	$f(x)$
-2	-8
-1	-2
-1/2	-1/2
0	0

inverse

x	$f(x)$

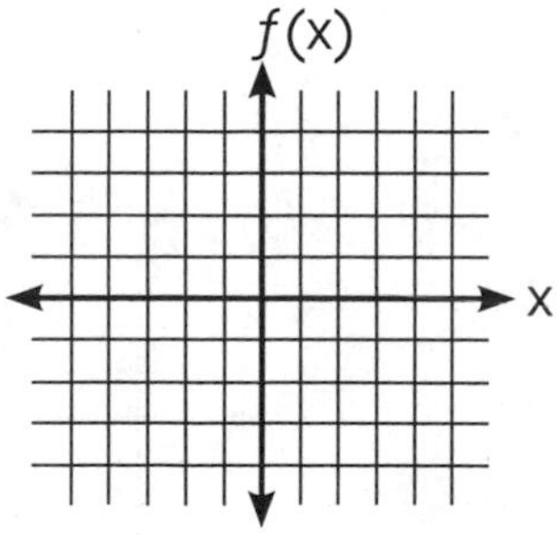

5.

function

x	$f(x)$
-2	8
-1	1
-1/2	1/8
0	0
1/2	-1/8
1	-1
2	-8

inverse

x	$f(x)$

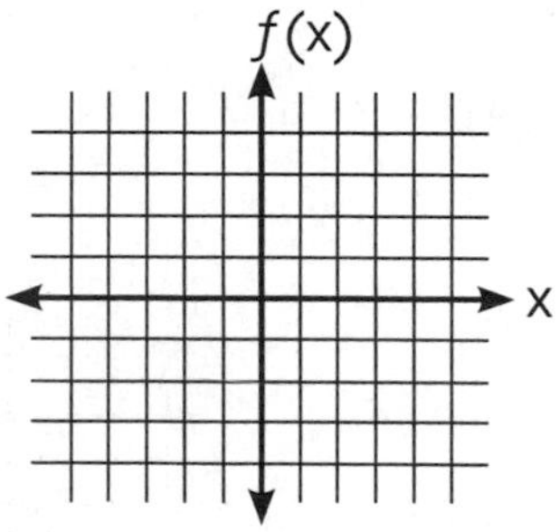

Name________________________________ HSF-BF.B.4c

More Graphing the Inverse of a Function

Find and graph the inverse of each function.

Example: $f(x) = 4x - 5$

$y = 4x - 5$

$x = 4y - 5$

$4y = x + 5$

$y = \frac{1}{4}x + \frac{5}{4}$

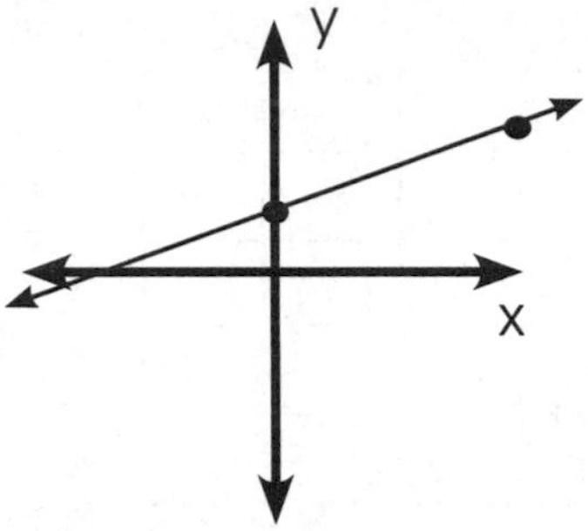

1. $f(x) = 2x^2$
(where $x \geq 0$)

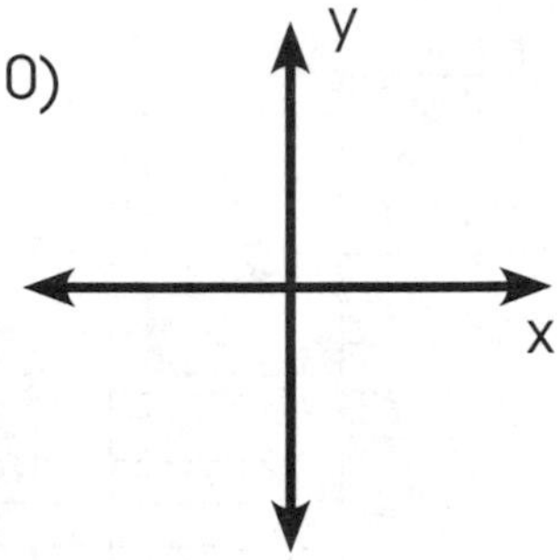

2. $f(x) = |x| + 2$
(where $x \geq 0$)

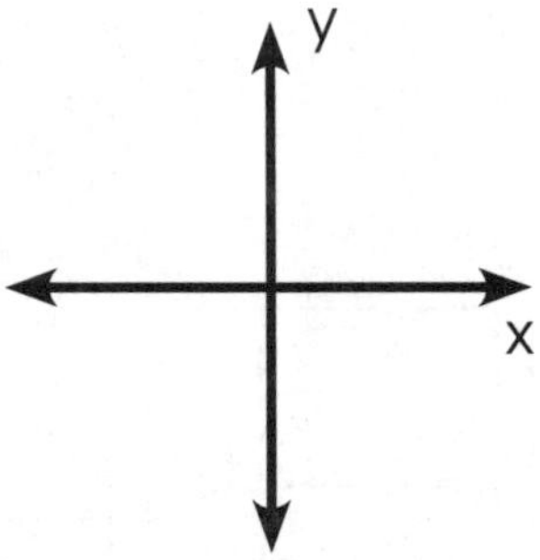

3. $f(x) = -x^2 + 1$
(where $x \geq 0$)

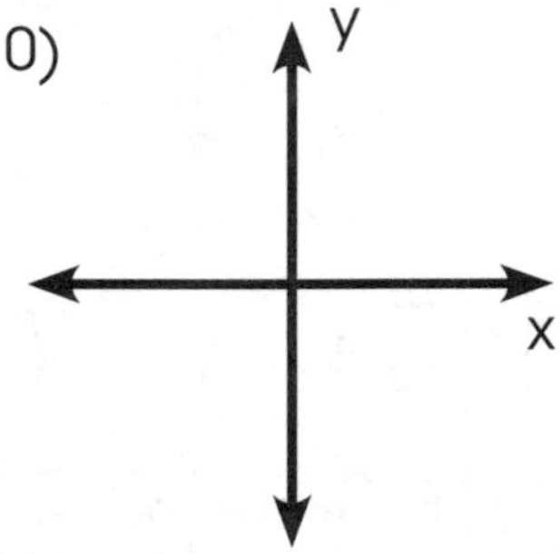

4. $f(x) = \frac{-1}{2}x^3$

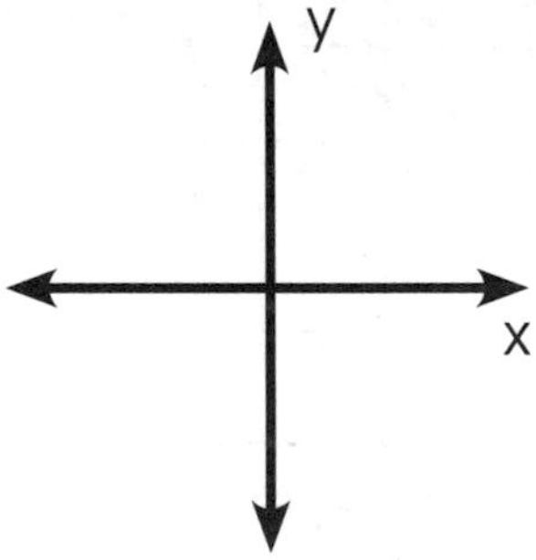

5. $f(x) = x - 1$
(where $x \geq 0$)

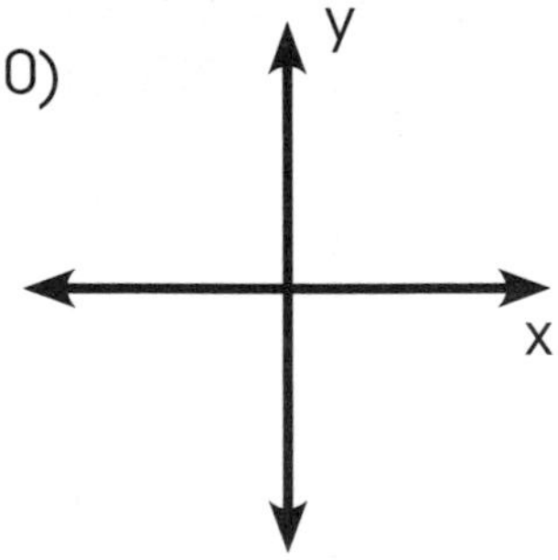

6. $f(x) = {}^{-}4x + 7$

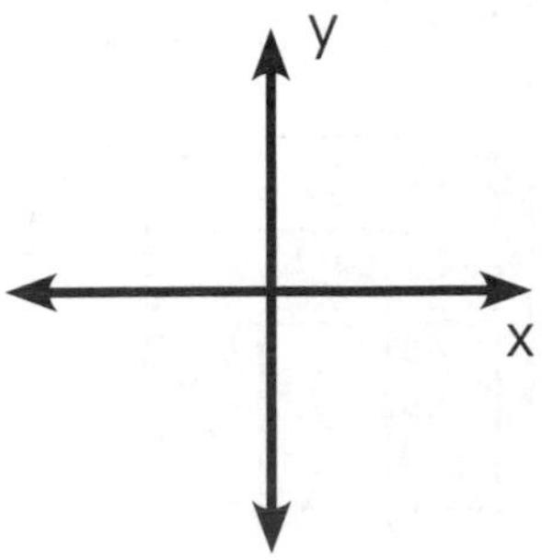

7. $f(x) = -x^2 + 5$
(where $x \geq 0$)

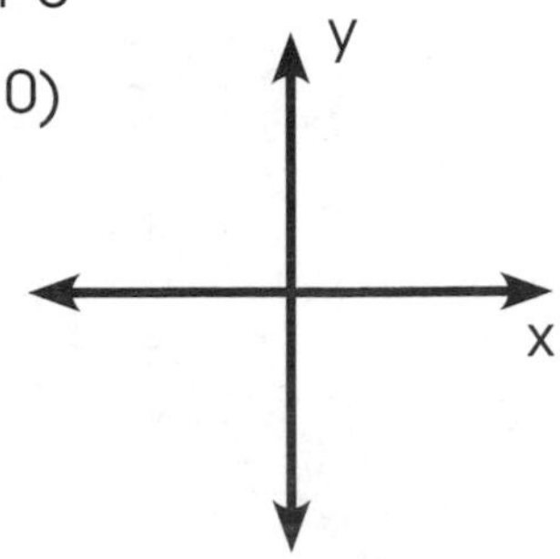

Name________________________________ HSA-REI.D.11, HSF-IF.C.7e

Graphing Exponential Functions

Graph each equation using a table of values. Include *x* values which are both negative and positive. Graph the equations on a separate piece of graph paper.

Example: $y = 2^x$

x	y
-1	1/2
0	1
1	2
2	4

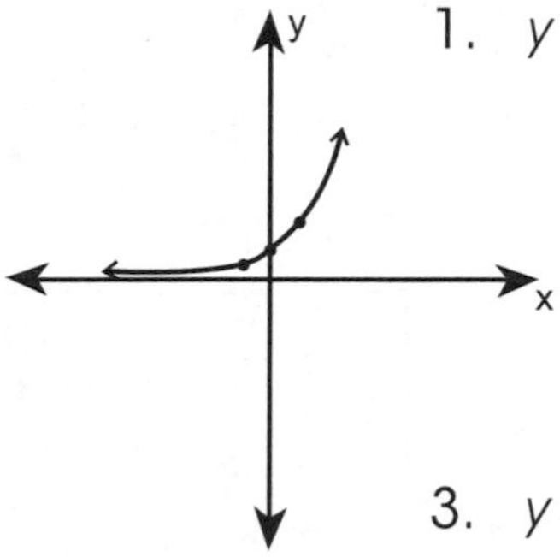

1. $y = 3^x$

x	y

2. $y = 10 \cdot 3^x$

x	y

3. $y = 10 \cdot 2^x$

x	y

4. $y = \left(\frac{1}{2}\right)^x$

x	y

5. $y = \left(\frac{1}{5}\right)^x$

x	y

6. $y = 50 \cdot 4^x$

x	y

7. $y = 10^x$

x	y

8. $y = \left(\frac{1}{10}\right)^x$

x	y

9. $y = 50 \cdot 2^x$

x	y

What do each of these sets of functions have in common?

Problems 1, 4, 5, 7, and 8:

Problems 4, 7, and 8:

Answer Key

1. Bill can paint a closet in 2 hours. Bob can paint the same closet in 3 hours. How long will it take them to paint the closet working together? $1\frac{1}{5}$ hrs.
2. Sally can address a box of envelopes in 30 minutes. Her brother Jim can address a box of envelopes in 1 hour. How long would it take both working together to address a box of envelopes? 20 min.
3. Paul can mow the grass in 50 minutes, but it takes Dan three times as long. How long will it take them to mow the grass if they work together? 37.5 min.
4. Using 1 drain, a swimming pool can be emptied in 45 minutes. Using a different drain, the job requires 1 hour and 15 minutes. How long will it take if both drains are opened? $28\frac{1}{8}$ min.
5. Susan can sort the office mail in 15 minutes; but if Kathy helps, they can sort the mail in 8 minutes. How long would it take Kathy to sort the mail alone? $17\frac{1}{7}$ min.
6. One pipe can fill a tank in 4 hours. A second pipe also requires 4 hours, but a third needs three hours. How long will it take to fill the tank if all three pipes are open? 1 hour and 12 min.

Page 7

1. How much simple interest can be earned in one year on $800 at 6%? $48
2. How long will it take $1,000 to double at 6% interest? 17 yrs.
3. Sam invested $1,600, part at 5% and the rest at 6%. The money earned $85 in one year. How much was invested at 5%? $1,100

Hint:

	P	x	r	x	t	=	I
Amount at 5%	x				1		
Amount at 6%	1,600 – x				1		

4. The Lewis family invested $900, part at 5% and the rest at 7%. The income from the investment was $58. How much was invested at 7%? $650
5. The Lockmores invested $7,000, part at 8% and part at $6\frac{1}{2}$%. If the annual return was $537.50, how much was invested at each rate? $5,500 at 8%, $1,500 at 6.5%
6. BDLV Associates had $7,400 invested at $5\frac{1}{2}$%. After part of the money was withdrawn, $242 was earned on the remaining funds for one year. How much money was withdrawn? $3,000
7. Michael has $2,000 more invested at $8\frac{1}{2}$% than he does at $9\frac{3}{4}$%. If the annual return from each investment is the same, how much is invested at each rate? $15,600 at 8.5%, $13,600 at 9.75%
8. Ms. Burke invested $53,650, part at 10.5% and the rest at 12%. If the income from the 10.5% investment is one third of that from the 12% investment, how much did she invest at each rate? $14,800 at 10.5%, $38,850 at 12%

Page 8

1. How much water must be added to 60kg of an 80% acid solution to produce a 50% solution? 36 kg of water
2. How much water must be evaporated from 8 grams of a 30% antiseptic solution to produce a 40% solution? 2 grams
3. How many grams of alcohol must be added to 40 grams of a 15% alcohol solution to obtain a 20% alcohol solution? 2.5 grams
4. How many quarts of antifreeze must be added to 15 quarts of a 30% antifreeze solution to obtain a 50% antifreeze solution? 6 quarts
5. A candy mixture is created with 2 types of candy, one costing $4 per pound and the other $3.50 per pound. How much of each type is needed for a 5 pound box that costs $18? 1 pound of $4 and 4 pounds of $3.50
6. A seed company mixes two types of seed for bird feeding. One costs $1.10 per kg and the other costs $2.25 per kg. How much of each type of seed is needed to produce 6kg at a cost of $8.90? 4 pounds of $1.10 and 2 pounds of $2.25
7. A farmer wants to mix milk containing 6% butterfat with 2 quarts of cream that is 15% butterfat to obtain a mixture that is 12% butterfat. How much milk containing 6% butterfat must he use? 1 quart
8. A store owner has 12 pounds of pasta worth 70¢ a pound. She wants to mix it with pasta worth 45¢ a pound so that the total mixture can be sold for 55¢ a pound (without any gain or loss). How much of the 45¢ pasta must she use? 18 pounds

Page 9

1. $P = \$100.00$ $r = 3\%$ $n = 4$ $t = 3$ years
$A = \$109.38$
2. $P = \$1,500.00$ $r = 4\%$ $n = 12$ $t = 1$ year
$A = \$1,561.11$
3. $P = \$2500.00$ $r = 5\%$ $n = 4$ $t = 10$ years
$A = \$4,109.05$
4a. $P = \$7500.00$ $r = 6\%$ $n = 1$ $t = 1$ year
$A = \$7950.00$
4b. $P = \$7900.00$ $r = 6\%$ $n = 4$ $t = 1$ year
$A = \$7963.73$
4c. In answer b, Juan is earning interest on interest.
5. $P = \$5,000.00$ $r = 6\%$ $n = 12$ $t = 18.5$ years
$A = \$15,129.89$

Page 10

Answer Key

C	The population of China is greater than 1,250,000,000.	1.25×10^{9}
E	Scientists at Oak Ridge National Laboratory have sent an electric current of 2,000,000 amperes/cm^2 down a wire.	2×10^{6}
E	The diameter of an electron is 0.0000000000011 cm.	1.1×10^{-12}
E	In an election in India, more than 343,350,000 people voted.	3.4335×10^{8}
E	The Earth's mass is 5,980,000,000,000,000,000,000 metric tons.	5.98×10^{21}
G	The Greenland–Canada boundary is about 1,700 miles long.	1.7×10^{3}
I	The isotope lithium 5 decays in 0.00000000000000000000044 seconds.	4.4×10^{-22}
I	The isotope tellurium 128 has a half-life of 1,500,000,000,000,000,000,000,000 years.	1.5×10^{24}
M	A microbe strain of H39 has a diameter of 0.0000003 m.	3×10^{-7}
N	A nugget of platinum found in 1843 weighed 340 ounces.	3.4×10^{2}
N	In 1996 the United States national debt was $5,129,000,000,000.	$\$5.129\times10^{12}$
O	A Saudi Arabia oil field contains about 82,000,000,000 barrels.	8.2×10^{10}
P	The fastest planet Mercury travels at 107,000 mph.	1.07×10^{5}
R	Sales of the record "White Christmas" exceeded 30,000,000.	3×10^{7}
V	In 1973, a vulture flying at 37,000 ft. collided with an aircraft.	3.7×10^{4}
X	The wavelength of an X ray is about 0.0000000015 m.	1.5×10^{-9}

What the decimal point said about scientific notation:

"It's a M O V I N G E X P E R I E N C E !"

M	O	V	I	N	G	E	X	P	E	R	I	E	N	C	E
-7	10	4	-22	2	3	6	-9	5	-12	7	24	8	12	9	21

Page 11

1. $x^{4+2} = x^6$
2. $x^{8-6} = x^2$
3. x^6y^3
4. $\frac{x^5}{y^{15}}$
5. $\frac{1}{y^{15}}$
6. x^{15}
7. a^{-3} or $\frac{1}{a^3}$
8. 2^3c^6 or $8c^6$
9. $\frac{n^{10}}{n^{10}} = 1$
10. $12a^8$
11. $\frac{v^4}{3^4}\cdot\frac{5^2}{v^2} = \frac{25v^2}{81}$
12. x^{-4} or $\frac{1}{x^4}$
13. $\frac{x}{2}$

Page 12

14. x^6y^{-3} or $\frac{x^6}{y^3}$
15. $4x^{-2}$ or $\frac{4}{x^2}$
16. $\frac{8d}{90d^{-2}}$ or $\frac{4d^3}{45}$
17. $\frac{^-2}{x^2}$
18. $x^{\frac{3}{3}}$ or x^1
19. $\left(\frac{125}{8x}\right)^2 = \frac{125^2}{64x^2}$ or $\frac{15625}{64x^2}$
20. $\frac{a^{13}b^6}{b^{-2}} = a^{13}b^8$
21. $\frac{1}{x^6y^{11}z}$
22. $\left(\frac{(xz)^2}{x^2}\right)^2 = \frac{(x^2z^2)^2}{x^4} = \frac{x^4z^4}{x^4} = z^4$
23. $\frac{a^{12}}{x^6y^3z^3b^{21}}$
24. $x^{-4}y^{-4}\cdot x^4y^{19} = y^{15}$
25. $\frac{x^{-12}}{y^{18}}\cdot\left(\frac{y^4}{x^4}\right) = x^{-16}y^{-14}$ or $\frac{1}{x^{16}y^{34}}$
26. $a^{12}b^6c^{48}\cdot a^{-9}b^4x^1 = a^3b^{10}c^{48}x$
27. $\frac{x^{12}b^3}{4^{-3}}\cdot 2x^5 = 128x^{17}b^3$
28. $a^{-36}b^8c^{-4}\cdot\frac{a^3b^3}{x^3} = \frac{a^{-33}b^{11}c^{-4}}{x^3} = \frac{b^{11}}{a^{33}c^4x^3}$
29. $\frac{a^3b^5x^2y^{11}}{2^{29}c^{12}}$

Page 13

1. $\frac{4}{9}$
2. $\frac{729}{64}$
3. $\frac{1}{8}$
4. $\frac{1}{5}$
5. $\frac{64}{125}$
6. $\frac{64}{27}$
7. $\frac{6}{11}$
8. $\frac{4}{9}$
9. $\frac{49}{81}$
10. $\frac{729}{8}$
11. $\frac{49}{16}$
12. $\frac{1,419,857}{759,375}$

As a check, for each problem number substitute the answer.

$$1 \cdot 2 \cdot 3 \div 4 \cdot 5 \cdot 6 \div 7 \div 8 \div 9 \div 10 \cdot 11 = \frac{22}{25}$$

$$\frac{4}{9} \cdot \frac{729}{64} \cdot \frac{1}{8} \div \frac{1}{5} \cdot \frac{64}{125} \cdot \frac{64}{27} \div \frac{6}{11} \div \frac{4}{9} \div \frac{49}{81} \div \frac{729}{8} \cdot \frac{49}{16} = \frac{22}{25}$$

Page 14

Answer Key

1. $\sqrt{100} =$ 10
2. $\sqrt{75} =$ $5\sqrt{3}$
3. $-\sqrt{144a^2} =$ $^{-}12a$
4. $\sqrt{128x^3} =$ $8x\sqrt{2x}$
5. $2\sqrt{1000} =$ $20\sqrt{10}$
6. $\sqrt{15a^8b} =$ $a^4\sqrt{15b}$
7. $\sqrt{16c^2d^2} =$ $4cd$
8. $2\sqrt{27x^5y} =$ $6x^2\sqrt{3xy}$
9. $-\sqrt{20xy^2} =$ $^{-}2y\sqrt{5x}$
10. $\sqrt{50a^3} =$ $5a\sqrt{2a}$
11. $\sqrt{96bc^2d^5} =$ $4cd^2\sqrt{6bd}$
12. $^{-}3\sqrt{150a^7c^2} =$ $^{-}15a^3c\sqrt{6a}$
13. $\sqrt{27a^2} =$ $3a\sqrt{3}$
14. $2\sqrt{50x^2yz^3} =$ $10xz\sqrt{2yz}$
15. $\sqrt{243m^5n^2} =$ $9m^2n\sqrt{3m}$
16. $-\sqrt{320y^9z^{10}} =$ $^{-}8y^4z^5\sqrt{5y}$

Page 15

1. $7mt\sqrt{t}$ **2.** $3x$ **3.** $8ab^2$
4. $2x\sqrt[4]{xy^3}$ **5.** $3x^2y^3$ **6.** $xy^{10}\sqrt[10]{1000x^2}$
7. $7d^2$ **8.** $^{-}2x - 1$ **9.** $(x + 1)^2$
10. $x + 1$ **11.** $2x - 3$

Page 16

1. $\sqrt{2} \bullet \sqrt{8} = 4$ 4
2. $5\sqrt{5} \bullet 3\sqrt{14} = 15\sqrt{70}$ $15\sqrt{70}$
3. $\sqrt{5b} \bullet \sqrt{10b} = 2b\sqrt{5}$ $5b\sqrt{2}$
4. $a\sqrt{2x} \bullet x\sqrt{6x} = 2ax\sqrt{3x}$ $2ax^2\sqrt{3}$
5. $2m\sqrt{7mn} \bullet 3\sqrt{7m} = 42m^2\sqrt{n}$ $42m^2\sqrt{n}$
6. $^{-}5a\sqrt{2a^4b} \bullet 4b\sqrt{12a^3b^4}$ $= {}^{-}40ab\sqrt{6a^7b}$ $^{-}40a^4b^3\sqrt{6ab}$
7. $2\sqrt{5}(-\sqrt{3x}) = {}^{-}2\sqrt{15x}$ $^{-}2\sqrt{15x}$
8. $5\sqrt{6} \bullet 2\sqrt{2} = 30\sqrt{2}$ $20\sqrt{3}$
9. $\sqrt{x} \bullet \sqrt{9x} = 3x$ $3x$
10. $\sqrt{2x} \bullet \sqrt{10x^2y} = 5y\sqrt{2xy}$ $2x\sqrt{5xy}$
11. $4x\sqrt{5} \bullet \sqrt{8xy^2} = 8xy\sqrt{10x}$ $8xy\sqrt{10x}$
12. $2\sqrt{x^3} \bullet 4\sqrt{x} = 8x^4$ $8x^2$
13. $^{-}7\sqrt{3y} \bullet \sqrt{6y} = {}^{-}14y\sqrt{3}$ $-21y\sqrt{2}$
14. $\sqrt{xy} \bullet \sqrt{xy} = \sqrt{xy}$ xy
15. $\sqrt{x^3y^5} \bullet \sqrt{x^2y} = x^2y^3\sqrt{x}$ $x^2y^3\sqrt{x}$

Page 17

1. $\sqrt{\frac{2ab^2}{c^2d}} = \frac{b\sqrt{2ad}}{cd}$
2. $\sqrt{\frac{2x}{3y}} = \sqrt{\frac{6xy}{3y}}$
3. $\sqrt{\frac{19x^2}{32}} = \frac{x\sqrt{38}}{8}$
4. $\sqrt{\frac{4a^2b}{x^8y^7}} = \frac{2a\sqrt{by}}{x^4y^4}$
5. $x\sqrt{\frac{5d}{3x^2}} = \frac{\sqrt{15d}}{3}$
6. $\sqrt{\frac{7a^2}{8cd}} = \frac{a\sqrt{14cd}}{4cd}$
7. $\sqrt{\frac{n^2}{7}} = \frac{n\sqrt{7}}{7}$
8. $\sqrt{\frac{8}{25}} = \frac{2\sqrt{2}}{5}$
9. $\sqrt{\frac{3\sqrt{2}}{\sqrt{3}}} = \sqrt{6}$
10. $\sqrt{\frac{4x^2}{25}} = \frac{2x}{5}$
11. $\sqrt{\frac{11y^3}{9}} = \frac{y\sqrt{11y}}{3}$
12. $\sqrt{\frac{25}{3x}} = \frac{5\sqrt{3x}}{3x}$
13. $\sqrt{\frac{3}{6x^3}} = \frac{\sqrt{2x}}{2x^2}$
14. $\frac{\sqrt{8x^2y}}{\sqrt{2y}} = 2x$

Page 18

1. $3\sqrt{7} - 4\sqrt{7} + 2\sqrt{7} =$ $\sqrt{7}$
2. $4\sqrt{27} - 2\sqrt{48} + \sqrt{147} =$ $11\sqrt{3}$
3. $5\sqrt{3} - 4\sqrt{7} - 3\sqrt{3} + \sqrt{7} =$ $2\sqrt{3} - 3\sqrt{7}$
4. $5\sqrt{x} - 3\sqrt{x} + a\sqrt{x} =$ $(a + 2)\sqrt{x}$
5. $4\sqrt{\frac{1}{2}} + 2\sqrt{18} - 6\sqrt{\frac{2}{9}} =$ $6\sqrt{2}$
6. $\sqrt{63} - \sqrt{28} - \sqrt{7} = 0$
7. $6\sqrt{3} - 2\sqrt{75} + 4\sqrt{\frac{3}{16}} =$ $^{-}3\sqrt{3}$
8. $\sqrt{50} + \sqrt{98} - \sqrt{75} + \sqrt{27} =$ $12\sqrt{2} - 2\sqrt{3}$
9. $2x\sqrt{ab} - 2y\sqrt{ab} + 4x\sqrt{ab} =$ $(6x - 2y)\sqrt{ab}$ or $2(3x - y)\sqrt{ab}$
10. $2b\sqrt{3c} + b\sqrt{5c} + b\sqrt{3c} - 2b\sqrt{5c} =$ $3b\sqrt{3c} - b\sqrt{5c}$
11. $4\sqrt{c^3d^3} + 3cd\sqrt{4cd} - 2c\sqrt{9cd^3} =$ $4cd\sqrt{cd}$
12. $8\sqrt{12} - \sqrt{10}\sqrt{\frac{1}{5}} - 108 + \sqrt{125} =$ $^{-}108 + 16\sqrt{3} + 3\sqrt{5}$
13. $x\sqrt{4x} + \sqrt{x^3} = 3x\sqrt{x}$
14. $3x\sqrt{7} + \sqrt{28x^2} - \sqrt{63x^2} = 2x\sqrt{7}$

Page 19

Answer Key

1. How high is the flag pole?

   75 ft.

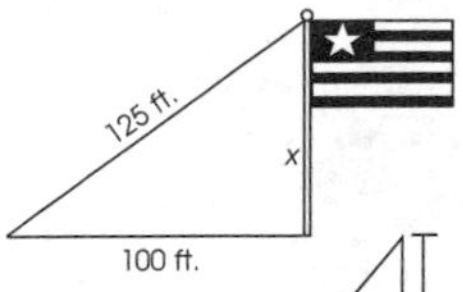

2. How long is the longest side of the sail?

   5.8 m

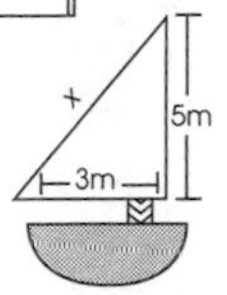

3. A 10-foot ladder is leaning against a house with its base 4 feet from the base of the house. How far up the house does the ladder touch the house? (Hint: Draw a picture.) 9.2 ft.
4. A 5-foot tall tree casts an 8-foot shadow on the ground. How far is it from the end of the shadow to the top of the tree? (Hint: Draw a picture.) 9.4 ft.
5. A guy wire is secured into the ground 15 feet from the base of a 36-foot pole. How long is the guy wire if it is attached at the top of the 36-foot pole? (Hint: Draw a picture.) 39 ft.
6. An airplane travels due east 65 miles, and then due north 72 miles. How far is the airplane from its starting point? (Hint: Draw a picture.) 97 miles

Page 20

Solve the problems to determine the oldest recorded ages for the given animals.

1. Black widow spider $\sqrt{5,000} \bullet \sqrt{2} =$ 100 days
2. Bedbug $\sqrt{10,000} + \sqrt{6,400} + \sqrt{4} =$ 182 days
3. Common housefly $\sqrt{400} - \sqrt{9} =$ 17 days
4. Queen Ant $\sqrt{\frac{1}{25}} \bullet \sqrt{8,100} =$ 18 years
5. Giant Centipede $\sqrt{36} + \sqrt{16} =$ 10 years
6. Goldfish $\sqrt{2,500} - \sqrt{81} =$ 41 years
7. Common toad $\sqrt{6,400} \div \sqrt{4} =$ 40 years
8. Boa constrictor $\sqrt{4} \bullet \sqrt{400} =$ 40 years
9. American alligator $\sqrt{36} \bullet \sqrt{121} =$ 66 years
10. Blue whale $\sqrt{2,500} - \sqrt{25} =$ 45 years
11. Octopus $\sqrt{\frac{1}{16}} \bullet \sqrt{144} =$ 3 years
12. Asian elephant $\sqrt{6,400} + \sqrt{1} =$ 81 years
13. Andean condor $\sqrt{144} \bullet \sqrt{36} =$ 72 years
14. Monarch butterfly $\sqrt{1.21} =$ 1.1 years
15. Giant salamander $10\sqrt{25} + \sqrt{1} =$ 51 years
16. Hedgehog $\sqrt{144} + \sqrt{4} =$ 14 years
17. Horse $2\sqrt{900} + \sqrt{4} =$ 62 years
18. Ostrich $5\sqrt{16} + 5\sqrt{100} - \sqrt{64} =$ 62 years
19. Sea anemone $\sqrt{25} \bullet 2\sqrt{81} =$ 90 years
20. Tortoise $\sqrt{22,500} + \sqrt{4} =$ 152 years

Page 22

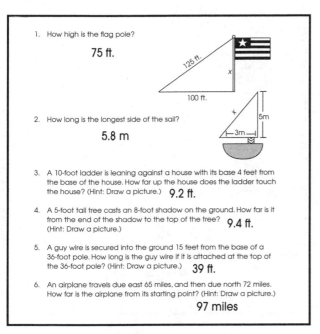

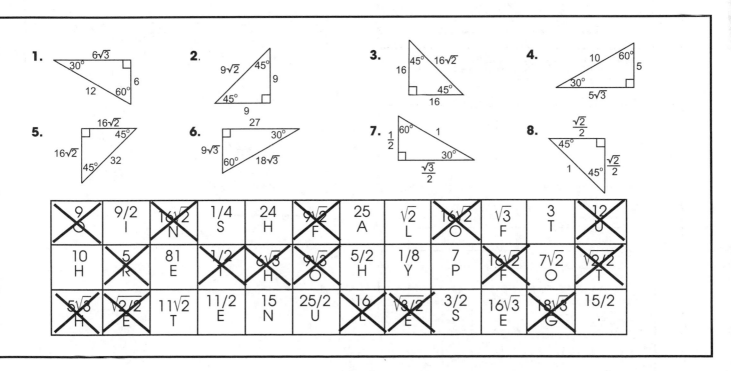

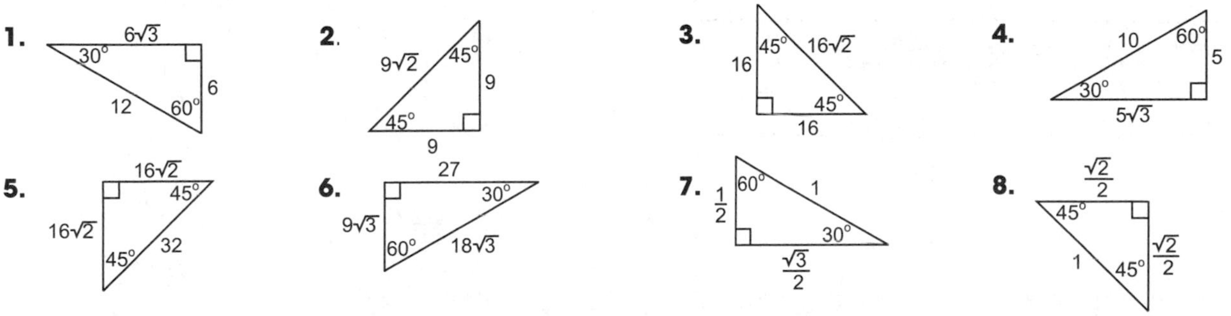

~~9 O~~	9/2 I	~~16√2 N~~	1/4 S	24 H	~~9√2 F~~	25 A	√2 L	~~16√2 O~~	√3 F	3 T	~~12 U~~
10 H	~~5 R~~	81 E	~~1/2 I~~	~~6√3 H~~	~~9√3 O~~	5/2 H	1/8 Y	7 P	~~16√2 F~~	7√2 O	~~√2/2 T~~
~~5√3 H~~	~~√2/2 E~~	11√2 T	11/2 E	15 N	25/2 U	~~16 L~~	~~√3/2 E~~	3/2 S	16√3 E	~~18√3 G~~	15/2 .

Page 21

Answer Key

1. $\sqrt{x-1}=4$	$\{17\}$	8. $4+\sqrt{x+1}=5$	$\{0\}$
2. $4=5\sqrt{x}$	$\left\{\frac{16}{25}\right\}$	9. $\frac{\sqrt{5-2x}}{3}=1$	$\{-2\}$
3. $\sqrt{x+3}=1$	$\{-2\}$	10. $\sqrt{4x-3}=\sqrt{x}$	$\{1\}$
4. $8=\sqrt{5a+}$	$\left\{\frac{63}{5}\right\}$	11. $5=\frac{15}{\sqrt{2a-3}}$	$\{6\}$
5. $2\sqrt{x}=5$	$\left\{\frac{25}{4}\right\}$	12. $6-\sqrt{y-5}=3$	$\{14\}$
6. $\sqrt{7+3x}=$	$\{3\}$	13. $2\sqrt{5}=3\sqrt{x}$	$\left\{\frac{20}{9}\right\}$
7. $\sqrt{4-x}=7$	$\{-45\}$	14. $2\sqrt{x}=4\sqrt{3}$	$\{12\}$

Page 23

1. 0.625 grams
2. 10 grams
3. 0.007 grams
4. 1031.25 grams
5. 6.8×10^{-11} grams
6. 22,920 years old

Page 24

Make your own matrix.

One week there is a birthday party every day. No two children are invited to the same party. Find out the day that each child attends a party. Start your matrix with Sunday and continue through Saturday.

1. Lisa and Pat don't go to a party on Friday or a Saturday.
2. Pat and Alice don't go on a Tuesday, but Sandy does.
3. Jennifer goes to a party on a Wednesday.
4. Jim goes to a party the day after Jennifer.
5. Lisa goes to a party the day before Pat.
6. Paul goes to a party on a Saturday.

	Sun.	Mon.	Tues.	Wed.	Thurs.	Fri.	Sat.
Lisa	X						
Pat		X					
Alice						X	
Sandy			X				
Jennifer				X			
Jim					X		
Paul							X

Page 25

1. $\frac{8a-8b}{a^2-b^2}=$	$\frac{8}{a+b}$	8. $\frac{2a^3+a^2-3a}{6a^3+5a^2-6a}=$	$\frac{a-1}{3a-2}$
2. $\frac{x^2+8x+16}{x^2-16}=$	$\frac{x+4}{x-4}$	9. $\frac{x^2-9}{x^2+x-6}=$	$\frac{x-3}{x-2}$
3. $\frac{12-4a}{a^2+a-12}=$	$\frac{-4}{a+4}$	10. $\frac{3x^2+2x-1}{x^2+3x+2}=$	$\frac{3x-1}{x+2}$
4. $\frac{t^2+4t-5}{t^2+9t+20}=$	$\frac{t-1}{t+4}$	11. $\frac{x^2+5x}{x^2-25}=$	$\frac{x}{x-5}$
5. $\frac{z^2-4z-5}{z^2+4z-45}=$	$\frac{z+1}{z+9}$	12. $\frac{a^2-11a+30}{a^2-9a+18}=$	$\frac{a-5}{a-3}$
6. $\frac{6b^3-24b^2}{b^2+b-20}=$	$\frac{6b^2}{b+5}$	13. $\frac{2y^3-12y^2+2y}{y^2-6y+1}=$	$2y$
7. $\frac{-x^2+8x-12}{x-2}=$	$6-x$	14. $\frac{a+b}{a^2+2ab+b^2}=$	$\frac{1}{a+b}$

Page 26

1. $\frac{24r^2s^2}{3s}\bullet\frac{-21s}{r}=-168rs^2$
2. $\frac{x^2y}{z^2}\bullet\frac{z}{xy}=\frac{x}{z}$
3. $\frac{2t+16}{4t}\bullet\frac{10t^2}{3t+24}=\frac{5}{3}t$
4. $\frac{x^2-1}{x}\bullet\frac{x^2}{x-1}=x^2+x$
5. $\frac{a+b}{a-b}\bullet\frac{a^2-b^2}{a+b}=a+b$
6. $\frac{a^2-4}{a^2-1}\bullet\frac{a-1}{a-1}=\frac{(a-2)(a+2)}{(a-1)(a+1)}$
7. $\frac{2x+2}{x-1}\bullet\frac{x^2+x-2}{x^2-x-2}=\frac{2(x+2)}{x-2}$
8. $\frac{z^2-6z-7}{z^2+z}\bullet\frac{z^2-z}{3z-21}=\frac{z-1}{3}$
9. $\frac{c^2-6c-16}{c^2+4c-21}\bullet\frac{c^2-8c+15}{c^2+9c+14}=\frac{(c-8)(c-5)}{(c+7)^2}$
10. $\frac{x+8}{x^2-x-12}\bullet\frac{x^2-6x+8}{x^2+6x-16}=\frac{1}{x+3}$
11. $\frac{h^2-2h-3}{h^2-9}\bullet\frac{h^2+5h+6}{h^2-1}=\frac{h+2}{h-1}$
12. $\frac{x^2-y^2}{x^2+4xy+3y^2}\bullet\frac{x^2+xy-6y^2}{x^2+xy-2y^2}=\frac{x-2y}{x+2y}$
13. $\frac{30+y-y^2}{25-y^2}\bullet\frac{y^2}{y^2-6y}\bullet\frac{y^2-y-12}{y^2-9}=\frac{y(y-4)}{(5-y)(y-3)}$
14. $\frac{5m+5n}{m^2-n^2}\bullet\frac{m^2-mn}{(m+n)^2}=\frac{5m}{(m+n)^2}$

Page 27

Answer Key

1. $\frac{b+2}{b^2-9} \div \frac{1}{b-3} = \frac{b+2}{b+3}$
2. $\frac{c^2+2cd}{2cd+d^2} \div \frac{c^3+2c^2d}{cd+d^2} = \frac{c+d}{c(2c+d)}$
3. $\frac{x^2+3x^3}{4-x^2} \div \frac{x+4x^2+3x^3}{2x+x^2} = \frac{x^2}{(2-x)(x+1)}$
4. $\frac{a^2-a-20}{a^2+7a+12} \div \frac{a^2-7a+10}{a^2+9a+18} = \frac{a+6}{a-2}$
5. $\frac{6a^2-a-2}{12a^2+5a-2} \div \frac{4a^2-1}{8a^2-6a+1} = \frac{3a-2}{3a+2}$
6. $\frac{a^3-6a^2+8a}{5a} \div \frac{2a-4}{10a-40} = (a-4)^2$
7. $\frac{12x+36}{x^2-2x-8} \div \frac{15x+45}{x^2+x-20} = \frac{4(x+5)}{5(x+2)}$
8. $\frac{x^2-y^2}{x^2+2xy+y^2} \div \frac{x-y}{x+y} = 1$
9. $(y^2-9) \div \frac{y^2+8y+15}{2y+10} = 2(y-3)$
10. $\frac{x^2-4x+4}{3x-6} \div (x-2) = \frac{1}{3}$
11. $\frac{(2a)^3}{(4bc)^3} \div \frac{16a^2}{8b^2c^3} = \frac{a}{16b}$
12. $\frac{\frac{26c^2}{5c^2d}}{\frac{13c^3}{25d^3}} = \frac{10d^2}{c^3}$

Page 28

1. $\frac{6x}{3x-7} \bullet \frac{9x-21}{21} \div \frac{x^2}{35} = \frac{30}{x}$
2. $\frac{x^2-x-6}{x^2+2x-15} \bullet \frac{x^2-25}{x^2-4x-5} \div \frac{x^2+5x+6}{x^2-1} = \frac{x-1}{x+3}$
3. $\frac{x-y}{x+y} \div \frac{5x^2-5y^2}{3x-3y} \bullet \frac{(x+y)^2}{x^2-y^2} = \frac{3}{5(x+y)}$
4. $(b^2-9) \div \frac{b^2+8b+15}{2b+10} \div (b-3) = 2$
5. $\frac{a^3b^3}{a^3-ab^2} \div \frac{abc}{a-b} \bullet \frac{ab+bc}{ab} = \frac{b^2(a+c)}{c(a+b)}$
6. $\frac{x^2+16x+64}{x^2-9} \div \frac{x^2-64}{x+3} \bullet (x^2-11x+24) = x+8$

Page 29

1. $\frac{2}{x} - \frac{8}{x} + \frac{3}{x} = \frac{^-3}{x}$
2. $\frac{3a}{5b} + \frac{2a}{5b} = \frac{a}{b}$
3. $\frac{r}{6} - \frac{5t}{6} = \frac{r-5t}{6}$
4. $\frac{x+y}{2} - \frac{x}{2} = \frac{y}{2}$
5. $\frac{c}{c-d} - \frac{d}{c-d} = 1$
6. $\frac{6a}{a+d} + \frac{6d}{a+d} = 6$
7. $\frac{x^2}{x-2} - \frac{4}{x-2} = x+2$
8. $\frac{c^2}{c^2-4} - \frac{6c+16}{c^2-4} = \frac{c-8}{c-2}$
9. $\frac{x^2-7x}{(x-3)^2} + \frac{12}{(x-3)^2} = \frac{x-4}{x-3}$
10. $\frac{x^2}{2x+14} - \frac{49}{2x+14} = \frac{x-7}{2}$
11. $\frac{7x}{2y+5} - \frac{6x}{2y+5} = \frac{x}{2y+5}$
12. $\frac{y+4}{y-5} - \frac{3y+1}{y-5} = \frac{-2y+3}{y-5}$
13. $\frac{2x-3}{2} - \frac{6x-5}{2} = {}^-2x+1$
14. $\frac{8a-1}{5} - \frac{3a-6}{5} = a+1$

Page 30

1. $\frac{1}{x} + \frac{1}{y} = \frac{y+x}{xy}$ or $\frac{x+y}{xy}$
2. $\frac{3n}{7} + \frac{n}{14} = \frac{n}{2}$
3. $\frac{2x}{3} + \frac{5y}{2} = \frac{4x+15y}{6}$
4. $\frac{x}{3} + \frac{x^2}{5} = \frac{3x^2+5x}{15}$
5. $\frac{2x}{x^2y} - \frac{y}{xy^2} = \frac{1}{xy}$
6. $\frac{5}{12xy} + \frac{3}{4x} = \frac{9y+5}{12xy}$
7. $\frac{a}{b} - \frac{c}{d} = \frac{ad-bc}{bd}$
8. $\frac{8}{x} + \frac{3}{xy} = \frac{8y+3}{xy}$
9. $\frac{4x-1}{3x} + \frac{x-8}{5x} = \frac{23x-29}{15x}$
10. $\frac{2x+1}{4} - \frac{x-1}{8} = \frac{3x+3}{8}$ or $\frac{3(x+1)}{8}$
11. $\frac{a+2b}{3} + \frac{a+b}{2} = \frac{5a+7b}{6}$
12. $\frac{1}{x} + \frac{2}{x^2} - \frac{3}{x^3} = \frac{(x+3)(x-1)}{x^3}$ or $\frac{x^2+2x-3}{x^3}$

Page 31

Answer Key

1. $\frac{3a+2b}{3b} - \frac{a+2b}{6a} = \frac{6a^2+3ab-2b^2}{6ab}$
2. $\frac{a}{2a+2b} - \frac{b}{3a+3b} = \frac{3a-2b}{6(a+b)}$
3. $\frac{3x}{2y-3} + \frac{2x}{3-2y} = \frac{x}{2y-3}$
   Hint: $3-2y = {}^-1\,(2y-3)$
4. $\frac{x}{x+3} + \frac{9x+18}{x^2+3x} = \frac{x+6}{x}$
5. $\frac{x+3}{x-5} + \frac{x-5}{x+3} = \frac{2(x^2-2x+17)}{(x-5)(x+3)}$
6. $\frac{11x}{x^2+3x-28} + \frac{x}{x+7} = \frac{x}{x-4}$
7. $\frac{d^2+3}{d^2-2d} - \frac{d-4}{d} = \frac{6d-5}{d(d-2)}$
8. $\frac{4a}{2a+6} - \frac{a-1}{a+3} = \frac{a+1}{a+3}$
9. $\frac{a+b}{ax+ay} - \frac{a+b}{bx+by} = \frac{(b+a)(b-a)}{ab(x+y)}$
10. $\frac{8}{c^2-4} + \frac{2}{c^2-5c+6} = \frac{10}{(c-3)(c+2)}$
11. $\frac{x}{x^2-16} + \frac{6}{4-x} - \frac{1}{x-4} = \frac{2(3x+14)}{(4-x)(x+4)}$
12. $\frac{1}{a^2-a-2} + \frac{1}{a^2+2a+1} = \frac{2a-1}{(a-2)(a+1)^2}$
13. $\frac{5}{3x-3} + \frac{x}{2x+2} - \frac{3x^2}{x^2-1} = \frac{{}^-15x^2+7x+10}{6(x+1)(x-1)}$
14. $\frac{x+1}{x^2-9} + \frac{4}{x+3} - \frac{x-1}{x-3} = \frac{-x^2+3x-8}{(x+3)(x-3)}$

Page 32

1. $b + \frac{6}{b-1} = \frac{(b-3)(b+2)}{b-1}$
2. $3 + \frac{a+2b}{a-b} = \frac{4a-b}{a-b}$
3. $x - y + \frac{1}{x+y} = \frac{x^2-y^2+1}{x+y}$
4. $7 + \frac{3}{a} + \frac{6}{b} = \frac{7ab+3b+6a}{ab}$
5. $\frac{5}{x+2} + 1 = \frac{x+7}{x+2}$
6. $d + 3 + \frac{2d-1}{d-2} = \frac{d^2+3d-7}{d-2}$
7. $\frac{2x-3}{x+2} - 4 = \frac{{}^-2x-11}{x+2}$
8. $2x - \frac{x+y}{y} = \frac{2xy-x-y}{y}$
9. $\frac{8}{3a-1} - 6 = \frac{14-18a}{3a-1}$
10. $(x-4) - \frac{1}{x-2} = \frac{x^2-6x+7}{x-2}$
11. $\frac{x}{2y} - (x+2) = \frac{x-2xy-4y}{2y}$
12. $(a+2) + \frac{7}{a-2} = \frac{a^2+3}{a-2}$
13. $4 - \frac{3}{y-1} - \frac{1}{y+1} = \frac{4y^2-4y-6}{y^2-1}$
14. $\frac{\frac{a}{b}+1}{\frac{a}{b}-1} = \frac{a+b}{a-b}$

Page 33

1. $y^2 - 13y + 36 \div y - 4 =$ $y - 9$
   Hint: $4 \,\big|\; 1 \quad {}^-13 \quad 36$
2. $x^2 + 10x + 21 \div x + 3 =$ $x + 7$
3. $4a^2 + 19a + 21 \div a + 1 =$ $4a + 15 + \frac{6}{a+1}$
4. $x^3 - 5x^2 + 2x + 8 \div x - 2 =$ $x^2 - 3x - 4$
5. $y^2 + 25 \div y + 5 =$ $y - 5 + \frac{50}{y+5}$
6. $x^3 + 2x^2 - 2x + 24 \div x + 4 =$ $x^2 - 2x + 6$

Page 34

1. $\frac{5}{6x} + 3 = \frac{1}{2x}$ $x = -\frac{1}{9}$
2. $\frac{2}{5n} = \frac{3}{10n} - \frac{3}{5}$ $n = -\frac{1}{6}$
3. $\frac{4}{3x} - \frac{5}{2x} = 5 + \frac{1}{6x}$ $x = -\frac{4}{15}$
4. $\frac{c-7}{c+2} = \frac{1}{4}$ $c = 10$
5. $\frac{y}{y-3} = 2$ $y = 6$
6. $\frac{2x}{5} + \frac{1}{2} = \frac{3x}{10}$ $x = {}^-5$
7. $\frac{x}{x-2} = \frac{4}{5}$ $x = {}^-8$
8. $\frac{2}{3} = \frac{y}{y+3}$ $y = 6$
9. $\frac{10}{x-3} = \frac{9}{x-5}$ $x = 23$
10. $\frac{7}{x} - \frac{4x}{2x-3} = {}^-2$ $x = 2\frac{5}{8}$
11. $\frac{3}{x} + \frac{1}{2x} = \frac{7}{8}$ $x = 4$
12. $\frac{3}{4} = \frac{x+5}{x-2}$ $x = {}^-26$

Page 35

Answer Key

1. $\frac{1}{u+4} + \frac{1}{u-4} = \frac{6}{u^2-16}$ — $u = 3$
2. $\frac{x}{8} + \frac{1}{x-2} = \frac{x+2}{2x-4}$ — $x = 0, x = 6$
3. $\frac{5y}{y+1} - \frac{y}{y+6} = 4$ — $y = 24$
4. $\frac{d}{d-2} = \frac{d+3}{d+2} - \frac{d}{d^2-4}$ — $d = {}^{-}3$
5. $\frac{6y}{2y+1} - \frac{3}{y} = {}^{-}1$ — $y = -\frac{3}{8}, y = 1$
6. $2 + \frac{4}{b-1} = \frac{4}{b^2-b}$ — $b = {}^{-}2$
7. $\frac{2z^2+z-3}{z^2+1} = 2$ — $z = 5$
8. $\frac{x}{x-3} + \frac{2}{x+4} = 1$ — $x = -\frac{6}{5}$
9. $\frac{1}{m-3} + \frac{1}{m+5} = \frac{m+1}{m-3}$ — $m = {}^{-}1, m = {}^{-}3$
10. $\frac{c}{c+1} + \frac{3}{c-3} + 1 = 0$ — $c = 0, c = 1$
11. $\frac{b}{b+1} - \frac{b+1}{b-4} = \frac{5}{b^2-3b-4}$ — No Solution
12. $\frac{2}{2y+1} - \frac{1}{2y} = \frac{3}{2y+1}$ — $y = -\frac{1}{4}$

Page 36

1. $(12 - 8) + 3 =$ 7
2. $2 \bullet 6 + 4 \bullet 5 =$ 32
3. $25 \div 5 \bullet 4 - 15 \bullet 8 =$ ${}^{-}100$
4. $3 + 15 \div 3 - 4 =$ 4
5. $15 \div (7 - 2) + 3 =$ 6
6. $2(7 + 3) \div 4 =$ 5
7. $7 - (8 \bullet 2) \bullet 0 =$ 7
8. $2(4 + (6 \div 2)) =$ 14
9. $20 \div (2 + (7 - 4)) =$ 4
10. $6({}^{-}9 + 4) \div 3 - 1 =$ ${}^{-}11$
11. $6 - 4(6 + 2) =$ ${}^{-}26$
12. $12 \div ((8 \div 2) \bullet (3 \div 3)) =$ 3
13. $\frac{9^2 - 11}{(3 + 4) \bullet 10} =$ 1
14. $\frac{3^2 - 4 \bullet 3 + 4}{9^2 - 4} =$ $\frac{1}{77}$
15. $\frac{3 \bullet 2 \div 6 + 2 \bullet 3 \div 6}{3^2 + 2^2 + 1^2} =$ $\frac{1}{7}$
16. $\frac{2 \bullet 4 - 6(2 + 1)}{1^2 - 3 \bullet 2} =$ 2
17. $\frac{(4 - 6)^2}{{}^{-}24 \div 12} =$ ${}^{-}2$
18. $\frac{2 \bullet 6 - (4 + 2)}{({}^{-}2 - 4 - 6) \div (2 - 1)} =$ $-\frac{1}{2}$
19. $\frac{{}^{-}3(4 - 9)}{35 \div {}^{-}7} =$ ${}^{-}3$
20. $3^5 \div 3^2 \div 3^2 \div 3 =$ 1

Page 37

1. $x + 7 = {}^{-}13$ — $x = {}^{-}20$
2. $x + 7 = 4$ — $x = {}^{-}3$
3. ${}^{-}14 + y = {}^{-}17$ — $y = {}^{-}3$
4. $y - 11 = 14$ — $y = 25$
5. $y - 5 = {}^{-}7$ — $y = {}^{-}2$
6. ${}^{-}20 + x = {}^{-}80$ — $x = {}^{-}60$
7. $6 + x = 29$ — $x = 23$
8. $a + 32 = {}^{-}4$ — $a = {}^{-}36$
9. ${}^{-}2 = x - 2$ — $x = 0$
10. ${}^{-}19 + y = 42$ — $y = 61$
11. $16 = z - 10$ — $z = 26$
12. $y + 73 = 0$ — $y = {}^{-}73$
13. ${}^{-}100 = b + ({}^{-}72)$ — $b = {}^{-}28$
14. $w - 5 = (8 - 13)$ — $w = 0$
15. $x + 2.5 = {}^{-}4.7$ — $x = {}^{-}7.2$
16. $a + 3.6 = {}^{-}0.2$ — $a = {}^{-}3.8$
17. $x - 6\frac{1}{4} = 12\frac{1}{2}$ — $x = 18\frac{3}{4}$
18. $2\frac{1}{5} + x = {}^{-}3\frac{1}{2}$ — $x = {}^{-}5\frac{7}{10}$
19. $n + \frac{1}{2} = \frac{3}{4}$ — $n = \frac{1}{4}$
20. $b - 1\frac{1}{3} = {}^{-}3\frac{5}{6}$ — $b = {}^{-}2\frac{1}{2}$

Page 38

1. $3x = {}^{-}21$ — $x = {}^{-}7$
2. ${}^{-}7y = 28$ — $y = {}^{-}4$
3. ${}^{-}28 = {}^{-}196x$ — $x = \frac{1}{7}$
4. ${}^{-}15a = {}^{-}45$ — $a = 3$
5. $-x = 17$ — $x = {}^{-}17$
6. ${}^{-}21 = {}^{-}2x$ — $x = 10\frac{1}{2}$
7. ${}^{-}12b = {}^{-}288$ — $b = 24$
8. $12x = {}^{-}60$ — $x = {}^{-}5$
9. $\frac{a}{5} = {}^{-}6$ — $a = {}^{-}30$
10. $-\frac{2}{5}y = {}^{-}14$ — $y = 35$
11. $\frac{3x}{4} = {}^{-}24$ — $x = {}^{-}32$
12. $-\frac{x}{3} = \frac{4}{9}$ — $x = {}^{-}1\frac{1}{3}$
13. $-\frac{3}{7} = \frac{a}{14}$ — $a = {}^{-}6$
14. $3a = -\frac{1}{4}$ — $a = -\frac{1}{12}$
15. $\frac{a}{2.4} = 0.26$ — $a = 0.624$
16. $-\frac{1}{99}y = 0$ — $y = 0$
17. ${}^{-}1.5x = 6$ — $x = {}^{-}4$
18. ${}^{-}12.5 = 4n$ — $n = {}^{-}3.125$
19. ${}^{-}3.7w = {}^{-}11.1$ — $w = 3$
20. $\frac{y}{6} = -\frac{2}{3}$ — $y = {}^{-}4$

Page 39

Answer Key

1. $5x - 3 = 22$ $x = 5$
2. $4a + 3 = {}^-5$ $a = {}^-2$
3. $5 - 7y = 33$ $y = {}^-4$
4. $5x - 11 = {}^-16$ $x = {}^-1$
5. ${}^-3 = 5x + 12$ $x = {}^-3$
6. $0 = 0.6x - 3.6$ $x = 6$
7. $6 - 8x = {}^-26$ $x = 4$
8. $5 = 5x + 27$ $x = {}^-4\frac{2}{5}$
9. $3(w + 3) = {}^-15$ $x = {}^-8$
10. $2(y + 1) - 5 = 7$ $y = 5$
11. $6 - \frac{2}{3}x = {}^-8$ $x = 21$
12. $0.3x - 4.2 = 2.7$ $x = 23$
13. $8.6 = 2.1 - 1.3y$ $y = {}^-5$
14. $5 - 4(y + 1) = {}^-3$ $y = 1$
15. ${}^-1 = \frac{-y}{4} - 6$ $y = {}^-20$
16. $\frac{5x}{6} + 34 = 9$ $x = {}^-30$
17. $\frac{{}^-2}{3}d + 3 = 11$ $d = {}^-12$
18. $1.2x + 6 = {}^-1.2$ $x = {}^-6$
19. $28 = \frac{17}{32}x - 23$ $x = 96$
20. $\frac{2x}{5} + 4 = {}^-12$ $x = {}^-40$

Page 40

1. $y - 16 - 3y = 0$
   $y = {}^-8$
2. $y - 4 - 4y = {}^-1$
   $y = {}^-1$
3. $8y + 4 - 2y = 22$
   $y = 3$
4. $12 = 3y - y + 4y$
   $y = 2$
5. $2(y - 3) - y = {}^-1$
   $y = 5$
6. $3y - 2(y + 4) = 8$
   $y = 16$
7. $2(y - 1) - y = {}^-2$
   $y = 0$
8. $5(2y - 2) + 4 = 4$
   $y = 1$
9. $3(y + 2) + 2(y - 3) = {}^-15$
   $y = {}^-3$
10. $4y - 2(y - 5) = {}^-2$
    $y = {}^-6$
11. $3y + 3(1 - y) - y = {}^-6$
    $y = 9$
12. $7y - 3(y + 6) = 10$
    $y = 7$
13. $\frac{6y}{8} + 12 = 84$
    $y = 96$
14. ${}^-2({}^-3 - 4y) = {}^-10$
    $y = {}^-2$
15. ${}^-6y + 9 + 4y = {}^-3$
    $y = 6$

ANSWER BANK

2 I	⁻2 T	4 A	1 Y	⁻7 Q	8 N	5 U	⁻5 L	⁻8 H	96 P
⁻8 O	⁻9 F	6 T	⁻3 I	⁻1 R	9 M	⁻4 B	⁻10 S	⁻6 R	10 W
15 C	⁻18 S	35 K	7 H	20 G	16 U	⁻16 E	3 P	⁻15 D	0 M

T R Y + U M P H ! = T R I U M P H

14 10 8   5 7 3 1 9   15 2 4 6 11 13 12

Page 41

1. $4x - 6 = x + 9$ $x = 5$
2. $4 - 7x = 1 - 6x$ $x = 3$
3. ${}^-4x - 3 = {}^-6x + 9$ $x = 6$
4. $41 - 2n = 2 + n$ $n = 13$
5. $6(2 + y) = 3(3 - y)$ $y = -\frac{1}{3}$
6. $4y = 2(y - 5) - 2$ $y = {}^-6$
7. $6x - 9x - 4 = {}^-2x - 2$ $x = {}^-2$
8. $-(x + 7) = {}^-6x + 8$ $x = 3$
9. $3 - 6a = 9 - 5a$ $a = {}^-6$
10. ${}^-9x + 6 = -x + 4$ $x = \frac{1}{4}$
11. $5x - 7 = {}^-10x + 8$ $x = 1$
12. $y + 3 = 4y - 18$ $y = 7$
13. ${}^-3(y + 3) = 2y + 3$ $y = {}^-2\frac{2}{5}$
14. $2({}^-3a + 5) = {}^-4(a + 4)$ $a = 13$
15. $7x - 3 = 2(x + 6)$ $x = 3$
16. ${}^-6x + 9 = 4(5 - x)$ $x = {}^-5\frac{1}{2}$
17. $3(x + 2) = {}^-5 - 2(x - 3)$ $x = {}^-1$
18. $2(x - 3) = (x - 1) + 7$ $x = 12$
19. $\frac{1}{3}(6y - 9) = {}^-2y + 13$ $y = 4$
20. $\frac{1}{6}(12 - 6x) = 5(x + 4)$ $x = {}^-3$

Page 42

1. $3(x - 5) = 21$
   $3x - 15 = 21$
   $3x = 36$
   $x = 12$
2. $x + 30 = 4x - 6$
   ${}^-3x + 30 = {}^-6$
   ${}^-3x = {}^-36$
   $x = 12$
3. ${}^-6 + 2x = 9 - 3x$
   ${}^-6 + 5x = 9$
   $5x = 15$
   $x = 3$
4. ${}^-6x + 9 = {}^-4x - 3$
   ${}^-2x + 9 = {}^-3$
   ${}^-2x = {}^-12$
   $x = 6$
5. $\frac{3x}{5} = \frac{2x}{5} + 10$
   $\frac{x}{5} = 10$
   $x = 50$
6. $\frac{3x}{8} + 8 = {}^-40$
   $\frac{3x}{8} = {}^-48$
   $x = {}^-128$

Page 43

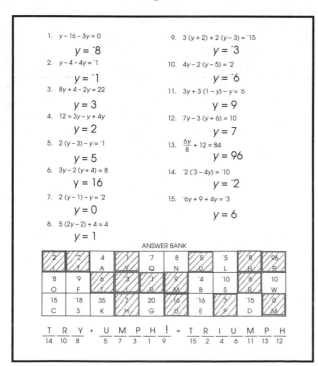

Answer Key

1. $\frac{3x}{2} - 9 = 0$ $x = 6$
2. $6x + 3 = {}^-5x + 14$ $x = 1$
3. $\frac{x}{8} + 3 = 2$ $x = {}^-8$
4. $5y = 2y - 42 - 3y$ $y = {}^-7$
5. $37 + 8x = 4(7 - x)$ $x = -\frac{3}{4}$
6. $5(2 - x) = 7x - 26$ $x = 3$
7. $6 + 4x = \frac{(6x + 9)}{3}$ $x = \frac{3}{2}$
8. $1.6(3y - 1) + 2 = 5y$ $y = 2$
9. $7x - 10 = 6(11 - 2x)$ $x = 4$
10. $3(4x - 9) = 5(2x - 5)$ $x = 1$
11. $(\frac{2}{3})(x + 9) = x + 5$ $x = 3$
12. $\frac{5y}{7} - 15 = 5y + 30$ $y = {}^-10\frac{1}{2}$

CHECK: Sum of the *X* solutions = $-\frac{1}{2}$ Sum of the *Y* solutions.

$7\frac{3}{4} = -\frac{1}{2}\ ({}^-15\frac{1}{2})$

Page 44

1. Twice a number, diminished by 17 is $^-3$. Find the number. 7
2. Six times a number, increased by 3 is 27. Find the number. 4
3. Three times the difference of 5 minus a number is 27. Find the number. $^-4$
4. Karl's team score is 39 points. This was one point less than twice Todd's team score. Find Todd's team score. 20
5. The length of a rectangle is 6 feet more than twice the width. If the length is 24 feet, what is the width? 9 feet
6. Four fifths of the third grade went on a trip to the zoo. If 64 children made the trip, how many children are in the third grade? 80
7. The price of a pack of gum today was 63¢. This is 3¢ more than three times the price ten years ago. What was the price ten years ago? 20¢
8. The sum of three consecutive integers is 279. Find the integers. 92, 93, 94
9. The sum of two consecutive odd integers is 112. Find the integers. 55, 57
10. Find four consecutive integers such that the sum of the second and fourth is 132. 64, 65, 66, 67
11. Find three consecutive odd integers such that their sum decreased by the second equals 50. 23, 25, 27

Page 45

1. One number is four times another. Their sum is 35. Find the numbers. 7, 28
2. The sum of two numbers is 21. One number is three less than the other. Find the numbers. 9, 12
3. The greater of two numbers is one less than 8 times the smaller. Their sum is 98. Find the numbers. 11, 87
4. In a triangle, the second angle measures twice the first, and the third angle measures 5 more than the second. If the sum of the angles' measures is 180°, find the measure of each angle. 35°, 70°, 75°
5. The length of a rectangle is 4 centimeters (cm) less than three times the width. The perimeter is 64 cm. Find the width and length. (Hint: Perimeter = 2l + 2w) w = 9, l = 23
6. The sum of three numbers is 64. The second number is 3 more than the first. The third number is 11 less than twice the first. Find the numbers. 18, 21, 25
7. Bill can type 19 words per minute faster than Bob. Their combined typing speed is 97 words per minute. Find Bob's typing speed. 39 words per minute

Page 46

$3x - 5 = 2(2x + 5)$ $^-15$	$10x + 8 = 12x - 18$ 13	$4(x - 7) = 2x - 6$ 11	$3(y + 4) = 5y + 30$ $^-9$
$3(2y + 4) = 4(y + 7) - 2$ 7	$^-6y = 10 - 4y$ $^-5$	$^-27 - 6x = 3x$ $^-3$	$7y + 3 = 12y - 2$ 1
$2y + 2 = 3y + 3$ $^-1$	$8x - (6x - 4) = 10$ 3	$6y = 10 + 4y$ 5	$6(x + 7) = 2(x + 7)$ $^-7$
Two consecutive whole numbers total 17. Find the larger. 9	Two consecutive odd numbers total $^-20$. Find the smaller. $^-11$	One number is 4 less than 3 times another. Their sum is $^-16$. Find the smaller. $^-13$	Two consecutive odd numbers total 32. Find the smaller. 15

Magic Sum is 0 .

Page 47

1. $^-8 - y = 22$ $^-30$
2. $18 = {}^-k + 3$ $^-15$
3. $4 - \frac{x}{5} = {}^-16$ 100
4. $-x - 15 = {}^-15$ 0
5. $-z = 11$ $^-11$
6. $^-28 = \frac{-y}{4} - 12$ 64
7. $^-82 = -a$ 82
8. $\frac{-b}{3} + 50 = 100$ $^-150$
9. $^-6 - x\frac{1}{9} = {}^-18$ 108
10. $^-3z + 5 = 38$ $^-11$
11. $-a\frac{1}{2} + 12 = {}^-9$ 42
12. $^-5y - {}^-7 = 52$ $^-9$

Page 48

Answer Key

1. $20y + 5 = 5y + 65$ — **4**
2. $13 - t = t - 7$ — **10**
3. $^-3k + 10 = k + 2$ — **2**
4. $^-9r = 20 + r$ — **$^-2$**
5. $6m - 2\frac{1}{2} = m + 12\frac{1}{2}$ — **3**
6. $18 + 4.5p = 6p + 12$ — **4**
7. $5x - \frac{1}{4} = 3x - \frac{5}{4}$ — **$-\frac{1}{2}$**
8. $-x - 2 = 1 - 2x$ — **3**
9. $3k + 10 = 2k - 21$ — **$^-31$**
10. $8y - 6 = 5y + 12$ — **6**
11. $-t + 10 = t + 4$ — **3**
12. $4m - 9 = 5m + 7$ — **$^-16$**

Page 49

1. A rectangle has a perimeter of 18 cm. Its length is 5 cm greater than its width. Find the dimensions. **2 x 7**
2. Timmy has 180 marbles, some plain and some colored. If there are 32 more plain marbles than colored marbles, how many colored marbles does he have? **74**
3. A theater sold 900 tickets to a play. Floor seats cost $12 each and balcony seats $10 each. Total receipts were $9,780. How many of each type of ticket were sold? **390 floor and 510 balcony**
4. Ryan and Karl spent 28 hours building a tree house. Ryan worked 4 more hours than Karl. How many hours did each work? **Karl 12 hrs, Ryan 16 hrs.**
5. The difference between seven times one number and three times a second number is 25. The sum of twice the first and five times the second is 95. Find the numbers. **10 and 15**
6. The sum of two numbers is 36. Their difference is 6. Find the numbers. **21 and 15**
7. The volleyball club has 41 members. There are 3 more boys than girls. How many girls are there? **19**
8. The sum of two numbers is 15. Twice one number equals 3 times the other. Find the numbers. **6 and 9**

Page 50

1. $y = 5x$ — Let $x = {}^-3, 0, 2$ — Note: This equation is already in the form of y = ..
   $y = 5x$ — $(^-3, {}^-15)$ $(0, 0)$ $(2, 10)$
2. $2x + y = 9$ — Let $x = {}^-1, 0, 5$
   $y = 9 - 2x$ — $(^-1, 11)$ $(0, 9)$ $(5, {}^-1)$
3. $-x = y + 3$ — Let $x = {}^-3, 0, 4$
   $y = -x - 3$ — $(^-3, 0)$ $(0, {}^-3)$ $(4, {}^-7)$
4. $y = \frac{2}{3}x + 1$ — Let $x = {}^-4, 0, 3$
   $y = \frac{2}{3}x + 1$ — $(^-4, -\frac{5}{3})$ $(0, 1)$ $(3, 3)$
5. $8x + y = 1$ — Let $x = {}^-2, 0, 1$
   $y = {}^-8x + 1$ — $(^-2, 17)$ $(0, 1)$ $(1, {}^-7)$
6. $y - 1 = {}^-3x$ — Let $x = {}^-3, 0, 2$
   $y = {}^-3x + 1$ — $(^-3, 10)$ $(0, 1)$ $(2, {}^-5)$
7. $2 = y - \frac{1}{3}x$ — Let $x = {}^-9, 0, 6$
   $y = \frac{1}{3}x + 2$ — $(^-9, {}^-1)$ $(0, 2)$ $(6, 4)$
8. $7x - y = {}^-8$ — Let $x = {}^-1, 0, {}^-3$
   $y = 7x + 8$ — $(^-1, 1)$ $(0, 8)$ $(^-3, {}^-13)$

Page 51

1. 12, 18 — **6**
2. 10, 35 — **5**
3. 8, 30 — **2**
4. 16, 24 — **8**
5. 28, 49 — **7**
6. 27, 63 — **9**
7. 30, 45 — **15**
8. 48, 72 — **24**

1. $6x + 3 =$ $3(2x + 1)$
2. $24x^2 - 8x =$ $8x(3x - 1)$
3. $6x - 12 =$ $6(x - 2)$
4. $2x^2 + 8x =$ $2x(x + 4)$
5. $4x + 10 =$ $2(2x + 5)$
6. $10x^2 + 35x =$ $5x(2x + 7)$
7. $10x^2y - 15xy^2 =$ $5xy(2x - 3y)$
8. $12x^2 - 9x + 15 =$ $3(4x^2 - 3x + 5)$
9. $3n^3 - 12n^2 - 30n =$ $3n(n^2 - 4n - 10)$
10. $9m^2 - 4n + 12 =$ prime
11. $2x^3 - 3x^2 + 5x =$ $x(2x^2 - 3x + 5)$
12. $13m + 26m^2 - 39m^3 =$ $13m(1 + 2m - 3m^2)$
13. $17x^2 + 34x + 51 =$ $17(x^2 + 2x + 3)$
14. $18m^2n^4 - 12m^2n^3 + 24m^2n^2 =$ $6m^2n^2(3n^2 - 2n + 4)$

Page 52

1. $x^2 - 1 =$ $(x + 1)(x - 1)$
2. $x^2 - 9 =$ $(x + 3)(x - 3)$
3. $x^2 + 4 =$ prime
4. $x^2 - 25 =$ $(x + 5)(x - 5)$
5. $9y^2 - 16 =$ $(3y + 4)(3y - 4)$
6. $4x^2 - 25 =$ $(2x + 5)(2x - 5)$
7. $9x^2 - 1 =$ $(3x + 1)(3x - 1)$
8. $a^2 - x^2 =$ $(a + x)(a - x)$
9. $25 - m^2 =$ $(5 + m)(5 - m)$
10. $x^2 - 16y^2 =$ $(x + 4y)(x - 4y)$
11. $25m^2 - n^2 =$ $(5m + n)(5m - n)$
12. $-x^2 + 16 =$ $(4 + x)(4 - x)$
13. $36m^2 - 121 =$ $(6m + 11)(6m - 11)$
14. $2x^2 - 8 =$ $2(x + 2)(x - 2)$
15. $25 + 4x^2 =$ prime
16. $4a^2 - 81b^2 =$ $(2a + 9b)(2a - 9b)$
17. $12x^2 - 75 =$ $3(2x + 5)(2x - 5)$
18. $a^2b - b^3 =$ $b(a + b)(a - b)$
19. $^-98 + 2x^2 =$ $2(x + 7)(x - 7)$
20. $5x^2 - 45y^2 =$ $5(x + 3y)(x - 3y)$
21. $9x^4 - 4 =$ $(3x^2 + 2)(3x^2 - 2)$
22. $16x^4 - y^2 =$ $(4x^2 + y)(4x^2 - y)$

Page 53

Answer Key

1. $a^2 - 36 =$
   $(a + 6)(a - 6)$
2. $9x^2 - 49 =$
   $(3x + 7)(3x - 7)$
3. $169m^2 - 4u^2 =$
   $(13m + 2u)(13m - 2u)$
4. $x^2y^2 - 9z^4 =$
   $(xy + 3z^2)(xy - 3z^2)$
5. $\frac{1}{4}x^2 - 25y^2 =$
   $(\frac{1}{2}x + 5y)(\frac{1}{2}x - 5y)$
6. $\frac{1}{9}x^2 - 16 =$
   $(\frac{1}{3}x + 4)(\frac{1}{3}x - 4)$
7. $64 - a^4b^4 =$
   $(8 + a^2b^2)(8 - a^2b^2)$
8. $y^6 - 100 =$
   $(y^3 + 10)(y^3 - 10)$
9. $\frac{4}{9}x^2y^2 - \frac{25}{36}z^2 =$
   $(\frac{2}{3}xy + \frac{5}{6}z)(\frac{2}{3}xy - \frac{5}{6}z)$
10. $y^8 - 81 =$
    $(y^4 + 9)(y^2 + 3)(y^2 - 3)$
11. $1 - 8u + 16u^2 =$
    $(4u - 1)^2$
12. $a^2b^2 + 6ab + 9 =$
    $(ab + 3)^2$
13. $x^2 + 2xy + y^2 =$
    $(x + y)^2$
14. $4x^2 + 12xy + 9y^2 =$
    $(2x + 3y)^2$
15. $100h^2 + 20h + 1 =$
    $(10h + 1)^2$
16. $9a^2 - 24a + 16 =$
    $(3a - 4)^2$
17. $4a^3 + 8a^2 + 4a =$
    $4a(a + 1)^2$
18. $5c + 20c^2 + 20c^3 =$
    $5c(2c + 1)^2$
19. $(x + 4)^2 - (y + 1)^2 =$
    $[(x + 4) + (y + 1)][(x + 4) - (y + 1)]$
20. $(x - 1)^2 - 10(x - 1) + 25 =$
    $[(x - 1) - 5]$ or $(x - 6)^2$

Page 54

1.	$3x^2 + 4x + 1 =$	T	A. $(7x - 4y)(4x + 3y)$
2.	$5x^2 + 7x + 2 =$	R	B. $(8x - 7y)(8x + 3y)$
3.	$2x^2 - 11x + 5 =$	N	C. $(7x + 8)(8x - 7)$
4.	$3x^2 + x - 2 =$	L	D. $(3x + 2)(4x + 3)$
5.	$5x^2 - 2x - 7 =$	O	E. $(2x - 3y)(x + 5y)$
6.	$8x^2 - 10xy + 3y^2 =$	S	H. Prime
7.	$6x^2 + 19x + 15 =$	I	I. $(2x + 3)(3x + 5)$
8.	$28x^2 + 5xy - 12y^2 =$	A	J. $(6x - 5)(3x - 7)$
9.	$2x^2 + 7xy - 15y^2 =$	E	L. $(3x - 2)(x + 1)$
10.	$12x^2 + 17x + 6 =$	D	N. $(2x - 1)(x - 5)$
11.	$4x^2 - 4xy - 5y^2 =$	H	O. $(5x - 7)(x + 1)$
12.	$56x^2 + 15y - 56 =$	C	P. $(4x - 7y)(3x - 2y)$
13.	$12x^2 - 29xy + 14y^2 =$	P	R. $(5x + 2)(x + 1)$
14.	$64x^2 - 32xy - 21y^2 =$	B	S. $(4x - 3y)(2x - y)$
15.	$16x^2 + 56xy + 49y^2 =$	U	T. $(3x + 1)(x + 1)$
16.	$18x^2 - 57x + 35 =$	J	U. $(4x + 7y)(4x + 7y)$

R E S I D E N T S I N T R A N S P A R E N T
2 9 6 7 10 9 3 1 6 7 3 1 2 8 3 6 13 8 2 9 3 1

A B O D E S S H O U L D N O T S E N D
8 14 5 10 9 6 6 11 5 15 4 10 3 5 1 6 9 3 10

P R O J E C T I L E S
13 2 5 16 9 12 1 7 4 9 6

Proverb: People in glass houses shouldn't throw stones.

Page 55

1. $2x^2 - 8 =$
   $2(x + 2)(x - 2)$
2. $2x^2 + 8x + 6 =$
   $2(x + 3)(x + 1)$
3. $3n^2 + 9n - 30 =$
   $3(n + 5)(n - 2)$
4. $6x^2 - 26x - 20 =$
   $2(3x + 2)(x - 5)$
5. $2x^2 + 12x - 80 =$
   $2(x + 10)(x - 4)$
6. $5t^2 + 15t + 10 =$
   $5(t + 1)(t + 2)$
7. $8n^2 - 18 =$
   $2(2n + 3)(2n - 3)$
8. $14x^2 + 7x - 21 =$
   $7(2x + 3)(x - 1)$
9. $4x^2 + 16x + 16 =$
   $4(x + 2)^2$
10. $18x + 12x^2 + 2x^3 =$
    $2x(x + 3)^2$
11. $2x - 2xy^2 =$
    $2x(1 + y)(1 - y)$
12. $3t^3 - 27t =$
    $3t(t + 3)(t - 3)$
13. $24a^2 \quad 30a + 9 =$
    $3(2a - 1)(4a - 3)$
14. $10x^2 + 15x - 10 =$
    $5(2x - 1)(x + 2)$
15. $3x^2 - 42x + 147 =$
    $3(x - 7)^2$
16. $4x^4 - 4x^2 =$
    $(4x^2)(x + 1)(x - 1)$

Page 56

1. $16x^2 - 40x - 24 =$
   $8(2x + 1)(x - 3)$
2. $27x^2 - 36x + 12 =$
   $3(3x - 2)^2$
3. $5x^2 - 60x - 140 =$
   $5(x - 14)(x + 2)$
4. $6m^3 + 54m^2 - 6m =$
   $6m(m^2 + 9m - 1)$
5. $5k^4 + 8k^3 - 4k^2 =$
   $k^2(5k - 2)(k + 2)$
6. $x^2y^4 - x^6 =$
   $x^2(y^2 + x^2)(y + x)(y - x)$
7. $y^4 - 6y^2 - 16 =$
   $(y^2 - 8)(y^2 + 2)$
8. $x^4 - 3x^2 - 4 =$
   $(x^2 + 1)(x + 2)(x - 2)$
9. $h^2 - (a^2 - 6a + 9) =$
   $[h + (a - 3)][h - (a - 3)]$
10. $81x^4 - 16y^4 =$
    $(9x^2 + 4y^2)(3x + 2y)(3x - 2y)$
11. $4mn^2 - 4m^2n^2 + m^3n^2 =$
    $mn^2(m - 2)^2$
12. $(2a + 3)^2 - (a - 1)^2 =$
    $[(2a + 3) + (a - 1)][(2a + 3) - (a - 1)]$
13. $16d^8 - 8d^4 + 1 =$
    $(2d^2 + 1)^2(2d^2 - 1)^2$
14. $x^2(x^2 - 4) + 4x(x^2 - 4) + 4(x^2 - 4) =$
    $(x + 2)^3(x - 2)$

Page 57

Answer Key

1. $x^2 - 5x - 6 = 0$ — $x = 6, {}^-1$
2. $v^3 - 4v = 0$ — $v = 0, {}^-2, 2$
3. $n^2 - 16n = 0$ — $n = 0, 16$
4. $x^2 + 9 = 10x$ — $x = 1, 9$
5. $6x^2 = 16x - 8$ — $x = \frac{2}{3}, 2$
6. $s^2 = 56s - s^3$ — $s = {}^-8, 0, 7$
7. $3y^2 + 2y - 1 = 0$ — $y = {}^-1, \frac{1}{3}$
8. $u^3 = 14u^2 + 32u$ — $u = {}^-2, 0, 16$
9. $23p = 5p^2 + 24$ — $p = \frac{8}{5}, 3$
10. $x^2 - 3x - 10 = 0$ — $x = {}^-2, 5$
11. $y^2 = 49$ — $y = {}^-7, 7$
12. $y^2 = {}^-7y - 10$ — $y = {}^-5, {}^-2$
13. $x^2 = 8x$ — $x = 0, 8$
14. $3x^2 - 2 = x^2 + 6$ — $x = {}^-2, 2$
15. $4y^2 = {}^-4y - 1$ — $y = -\frac{1}{2}$
16. $5x^2 - 2x - 3 = 0$ — $x = -\frac{3}{5}, 1$

Page 58

1. Fourteen less than the square of a number is the same as five times the number. Find the number. **7 or ${}^-2$**
2. When a number is added to six times its square, the result is 12. Find the number. **$-\frac{3}{2}$ or $\frac{4}{3}$**
3. Find two consecutive, negative integers whose product is 156. **${}^-13, {}^-12$**
4. The sum of the squares of two consecutive integers is 41. Find the integers. **${}^-5, {}^-4$ or 4, 5**
5. The sum of the squares of three consecutive, positive integers is equal to the sum of the squares of the next two integers. Find the five integers. **10, 11, 12, 13, 14**
6. Find two consecutive even integers whose product is 80. **8, 10**
7. Twice the square of a certain positive number is 144 more than twice the number. What is the number? **9**
8. The square of a positive number decreased by 10 is 2 more than 4 times the number. What is the number? **6**

Page 59

1. The length of a rectangle is 5m greater than twice its width, and its area is 33m². Find the dimensions. **3 x 11**
2. The perimeter of a rectangular piece of property is 8 miles, and its area is 3 square miles. Find the dimensions. (Hint: $\frac{1}{2}P = l + w$) **1 x 3**
3. When the dimensions of a 2cm x 5cm rectangle were increased by equal amounts, the area was increased by 18cm². Find the dimensions of the new rectangle. **4 x 7**
4. If the sides of a square are increased by 3 in., the area becomes 64 in.² Find the length of a side of the original square. **5**
5. A rug placed in a 10 ft. x 12 ft. room covers two-thirds of the floor area and leaves a uniform strip of bare floor around the edges. Find the dimensions of the rug. **8 x 10**
6. The area of a rectangular pool is 192 square meters. The length of the pool is 4 meters more than its width. Find the length and the width. **12 x 16**

Page 60

1. $x^2 + 2x + xy + 2y =$ $(x + 2)(x + y)$
2. $3a^2 - 2b - 6a + ab =$ $(a - 2)(3a + b)$
3. $t^3 - t^2 + t - 1 =$ (Hint: $t - 1 = 1(t - 1)$) $(t - 1)(t^2 + 1)$
4. $10 + 2t - 5s - st =$ $(2 - s)(t + 5)$
5. $\frac{2}{3}bc - \frac{14}{3}b + c - 7 =$ $(c - 7)(\frac{2}{3}b + 1)$
6. $4u^2 + v + 2uv + 2u =$ $(2u + 1)(2u + v)$
7. $ad + 3a - d^2 - 3d =$ $(a - d)(d + 3)$
8. $n^2 + 2n + 3mn + 6m =$ $(n + 3m)(n + 2)$
9. $2ax^2 + bx^2 - 2ay^2 - by^2 =$ $(x - y)(x + y)(2a + b)$
10. $yz^2 - y^3 + z^3 - y^2z =$ $(2 - y)(2 + y)^2$
11. $y^3 - y^2 - 4y + 4 =$ $(y + 2)(y - 2)(y - 1)$
12. $x^2a + x^2b - 16a - 16b =$ $(x + 4)(x - 4)(a + b)$
13. $x^3 + x^2 - x - 1 =$ $(x + 1)^2(x - 1)$
14. $a^3 - a^2 - 8a + 8 =$ $(a - 1)(a^2 - 8)$

Page 61

1. $x = 6, 4$
   1. $x - 6 = 0, x - 4 = 0$
   2. $(x - 6)(x - 4) = 0$
   3. $x^2 - 4x - 6x + 24 = 0$
   4. $x^2 - 10x + 24 = 0$
2. $x = {}^-7, {}^-2$
   1. $x + 7 = 0, x + 2 = 0$
   2. $(x + 7)(x + 2) = 0$
   3. $x^2 + 2x + 7x + 14 = 0$
   4. $x^2 + 9x + 14 = 0$
3. $x = 1, 5$
   1. $(x - 1)(x - 5)$
   2. $(x - 1)(x - 5) = 0$
   3. $x^2 - 5x - x + 5 = 0$
   4. $x^2 - 6x + 5 = 0$
4. $x = \frac{1}{2}, 3$
   1. $2x - 1 = 0, x - 3 = 0$
   2. $(2x - 1)(x - 3) = 0$
   3. $2x^2 - 6x - x + 3 = 0$
   4. $2x^2 - 7x + 3 = 0$
5. $x = -\frac{3}{4}, 2$
   1. $4x + 3 = 0, x - 2 = 0$
   2. $(4x + 3)(x - 2) = 0$
   3. $4x^2 - 8x + 3x - 6 = 0$
   4. $4x^2 - 5x - 6 = 0$
6. $x = -\frac{1}{4}, {}^-8$
   1. $4x + 1 = 0, x + 8 = 0$
   2. $(4x + 1)(x + 8) = 0$
   3. $4x^2 + 32x + x + 8 = 0$
   4. $4x^2 + 33x + 8 = 0$
7. $x = 1, {}^-1$
   1. $x - 1 = 0, x + 1 = 0$
   2. $(x - 1)(x + 1) = 0$
   3. $x^2 + x - x - 1 = 0$
   4. $x^2 - 1 = 0$
8. $x = 0, 9$
   1. $x = 0, x - 9 = 0$
   2. $x(x - 9) = 0$
   3. $x^2 - 9x = 0$
   4. $x^2 - 9x = 0$
9. $x = 0, {}^-2, {}^-3$
   1. $x = 0, x + 2 = 0, x + 3 = 0$
   2. $x(x + 2)(x + 3) = 0$
   3. $x^3 + 3x^2 + 2x^2 + 6 = 0$
   4. $x^2 + 5x^2 + 6x = 0$

Three solutions mean three factors.

$4x^2 - 5x - 6 = 0$ [illegible]	$x^2 - 10x + 24 = 0$ A	$x^2 - 6x + 5 = 0$ [illegible]	$x^3 + 5x^2 + 6x = 0$ [illegible]	$x^2 - 9x - 14 = 0$ C
$x^2 - 1 = 0$ L	$4x^2 + 33x + 8 = 0$ M	$4x^2 + 5x + 6 = 0$ U	$x^2 - 9x = 0$ N	$2x^2 + x + 3 = 0$ B
$x^2 + x - 1 = 0$ I	$x^2 + 9x + 14 = 0$ E	$x^2 - 2x - 24 = 0$ A	$x^2 + 1 = 0$ C	$2x^2 - 7x + 3 = 0$ R

The remaining letters spell the name of the type of equation in problem 9: **A CUBIC**

Page 62

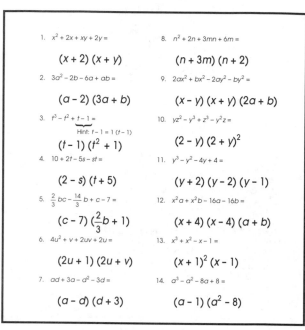

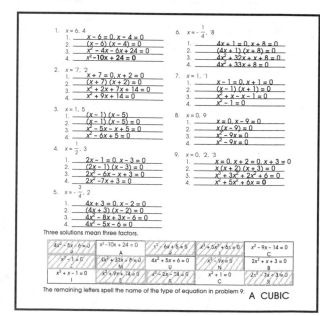

Answer Key

1. Maureen always finds things for both her collections outdoors.
2. Joan's friend enjoys collecting stamps.
3. One of Bryan's friends enjoys collecting coins.
4. The person who collects comics does not collect baseball cards.
5. One of Bryan's hobbies involves lots of reading.
6. Joan's family has a beach house; this is very helpful for one of her colloctions.
7. One of the girls collects dolls.

	Seashells	Stamps	Baseball Cards	Coins	Comic Books	Dolls	Bugs	Rocks
Maureen							X	X
Joan	X					X		
Robert			X	X				
Bryan		X			X			

Page 63

1. $y = \frac{1}{2}x - 3$

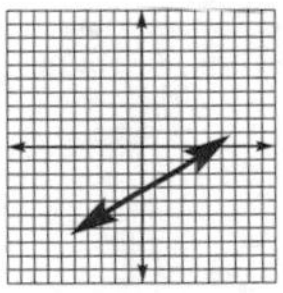

2. $^{-}2x + y = 5$

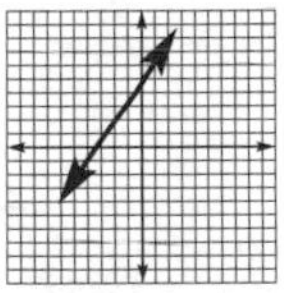

3. $4x + y = {}^{-}7$

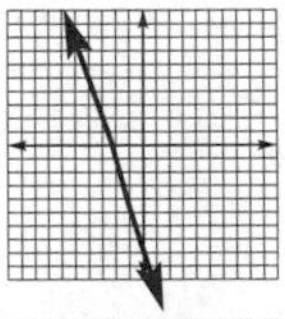

4. $y - 3 = 2x$

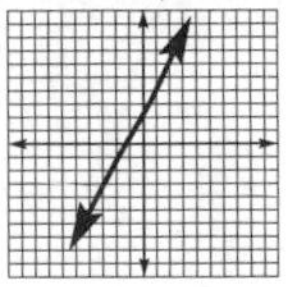

5. $3x + y = 7$

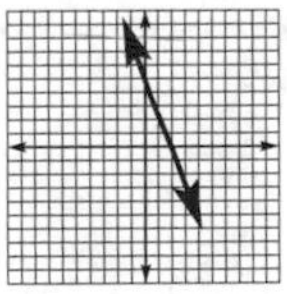

6. $3x - y = {}^{-}2$

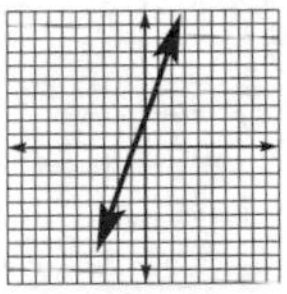

7. $y + 7 = 5x$

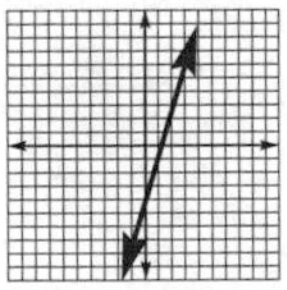

8. $y = \frac{1}{4}x - 2$

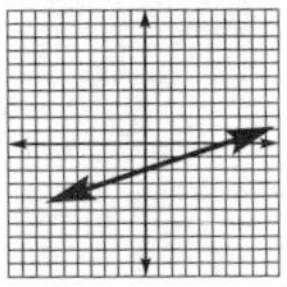

Page 65

9. $\frac{3}{4}x + y = 2$

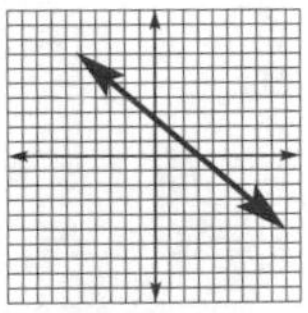

10. $x - y = {}^{-}4$

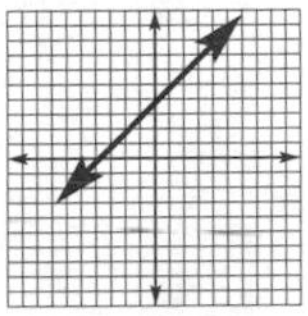

11. $^{-}3x + y = {}^{-}4$

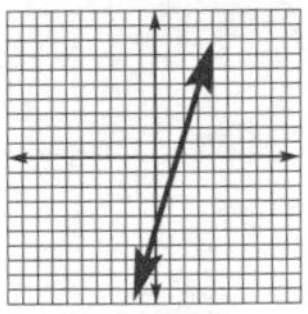

12. $y - x = {}^{-}1$

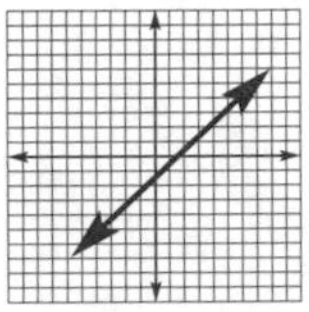

13. $y = -\frac{1}{2}x$

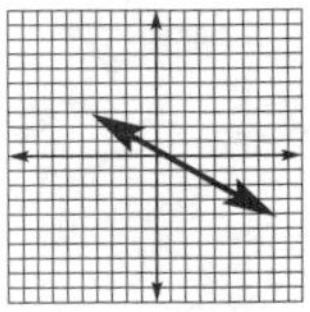

14. $\frac{1}{4}x + y = {}^{-}2$

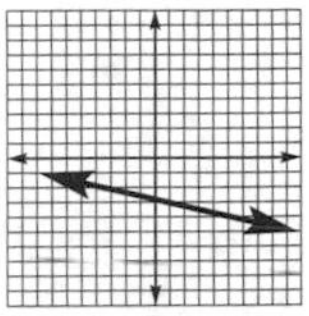

15. $y = {}^{-}4x - 5$

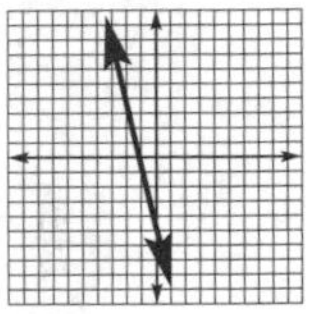

16. $y = {}^{-}2x - 3$

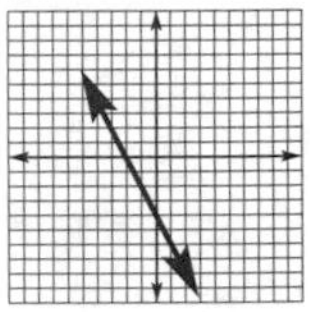

Page 66

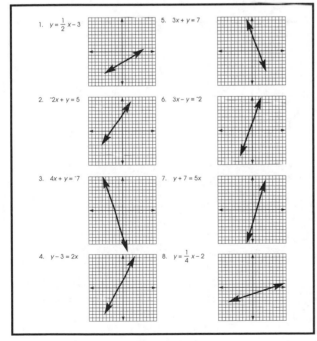

Answer Key

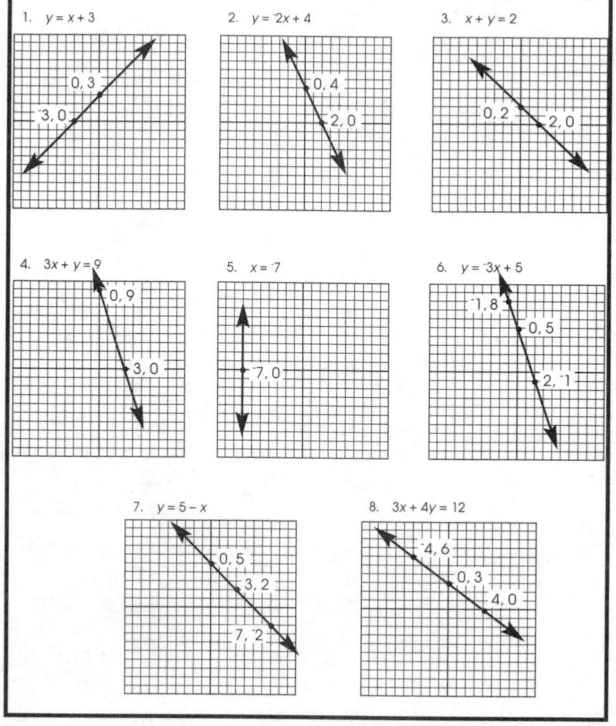

Page 67

1. $3x + 4y = 12$ — 1. $x_0 = 4, 0$ $y_0 = (0, 3)$
2. $4x + y = 2$ — 2. $x_0 = \frac{1}{2}$ $y_0 = (0, 2)$
3. $5x - 4y = 15$ — 3. $x_0 = (3, 0)$ $y_0 = (0, {}^-3\frac{3}{4})$
4. $2x - 2y = {}^-4$ — 4. $x_0 = (-2, 0)$ $y_0 = (0, 2)$
5. $3x + y = {}^-9$ — 5. $x_0 = ({}^-3, 0)$ $y_0 = (0, {}^-9)$
6. $4x - 2y - 8 = 0$ — 6. $x_0 = (2, 0)$ $y_0 = (0, {}^-4)$

II. Find the *x* and *y* intercepts. Then graph.

$x_0 = 5$
$y_0 = 2.5$

$x_0 = 0$
$y_0 = 0$

$x_0 = -\frac{1}{2}$
$y_0 = \frac{2}{3}$

$x_0 = 2$
$y_0 = 3$

$x_0 = \frac{12}{5}$
$y_0 = -\frac{12}{7}$

$x_0 = \frac{25}{4}$
$y_0 = 5$

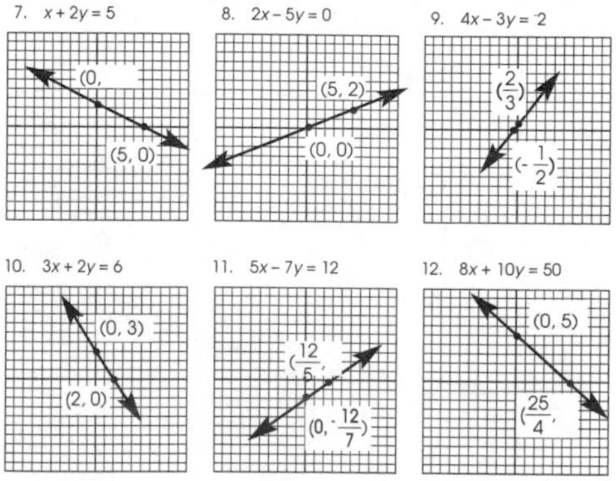

Page 68

1.

x	y
$^-$1	9
0	6
1	3

2.

x	y
$^-$1	$^-$3
0	$^-$2
1	$^-$1

3.

x	y
$^-$1	0
0	$^-$2
1	$^-$4

4.

x	y
$^-$1	11/2
0	5/2
1	$^-$1/2

5.

x	y
$^-$3	1/2
0	3/2
3	5/2

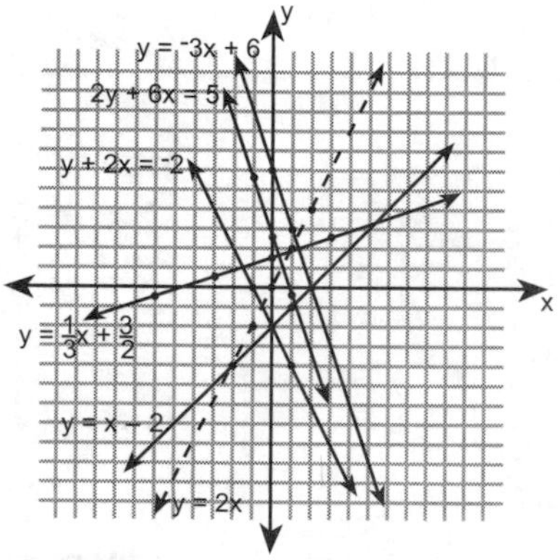

Page 69

1.

x	y
0	$^-$6
4	0

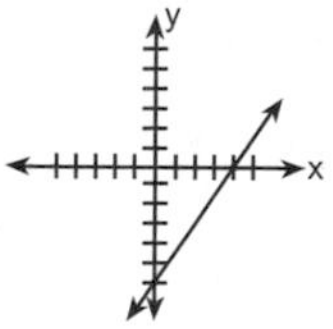

2.

x	y
0	5
$^-$5	0

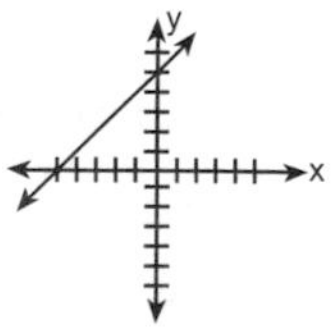

3.

x	y
0	$^-$10
2	0

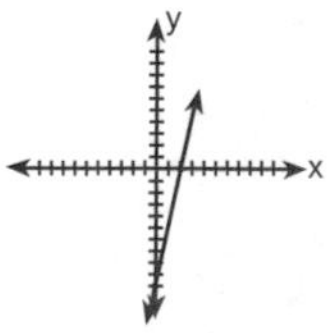

4.

x	y
0	5
10	0

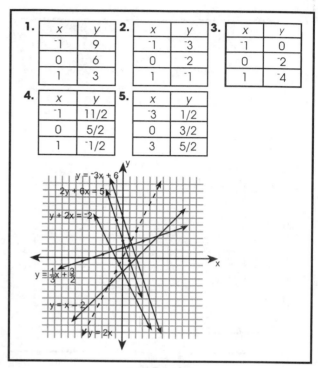

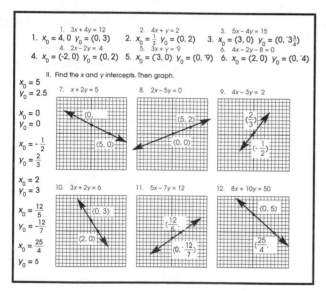

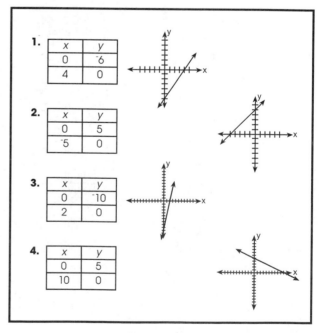

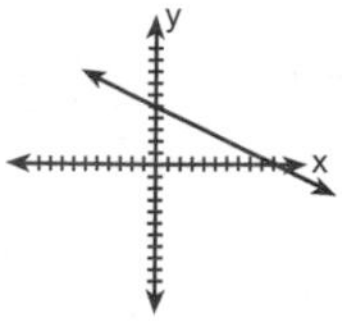

Page 70

Answer Key

5.

x	y
0	5
-1	0

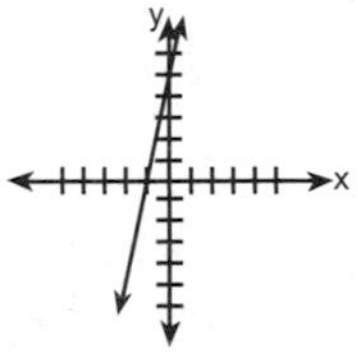

6.

x	y
0	-6
2	0

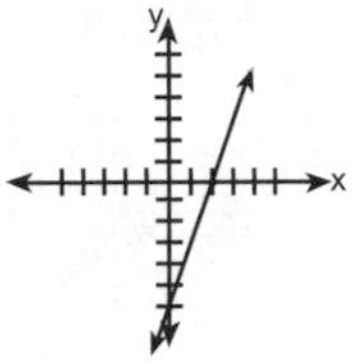

7.

x	y
0	-2
-3	0

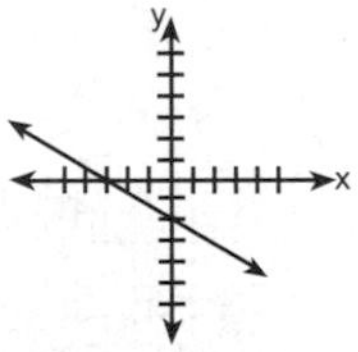

8.

x	y
0	-8
1/2	0

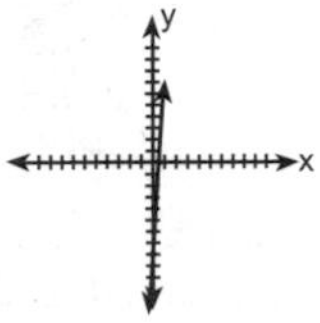

9.

x	y
0	12
9	0

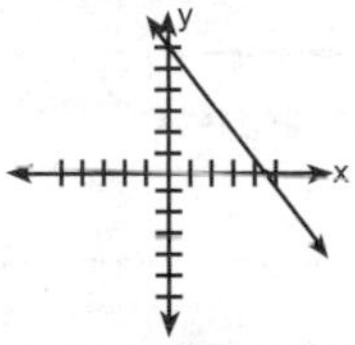

Page 70

1.
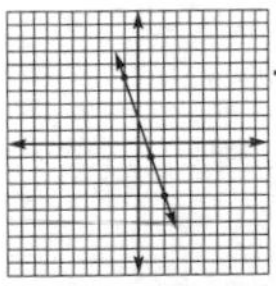
-3

3.
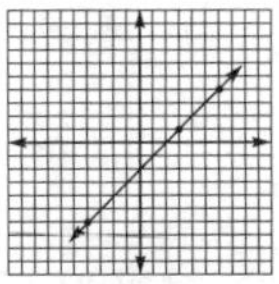
1

5.
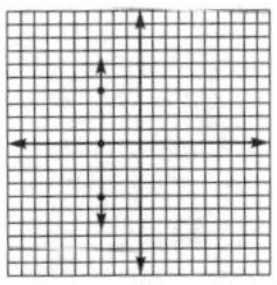
no slope

2.
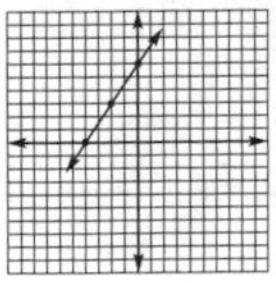
$\frac{3}{2}$

4.
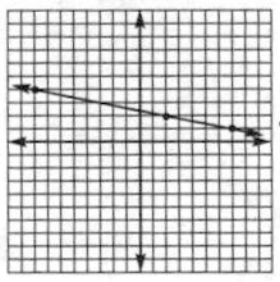
$-\frac{1}{5}$

6.
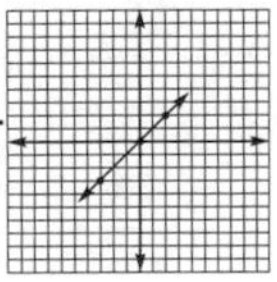
1

Page 71

1. A (-3, 1) D (4, 5) $\frac{4}{7}$
2. C (2, 6) F (3, 5) -1
3. B (0, 8) G (3, 2) -2
4. J (-6, -3) K (-4, 5) 4
5. P (9, 4) M (7, 3) $\frac{1}{2}$
6. Q (0, -4) R (1, -6) -2
7. L (-2, 6) N (2, -3) $-\frac{9}{4}$
8. S (-1, -3) X (2, -6) $\frac{-3}{2}$
9. T (-4, -4) 7 (6, 3) $\frac{7}{10}$
10. V $(\frac{3}{4}, \frac{3}{2})$ W $(\frac{11}{4}, \frac{5}{2})$ $\frac{1}{2}$
11. U (2, 3) A (-2, 3) 0
12. C (4, -1) D (-2, 2) $-\frac{1}{2}$
13. Z (3, 5) H (5, 10) $\frac{5}{2}$
14. J (-2, -3) K (13, 7) $\frac{2}{3}$

Page 72

1. (-5, -2), $-\frac{1}{2}$
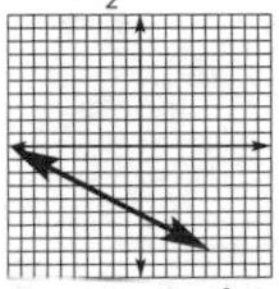

3. (2, -3), $\frac{3}{4}$
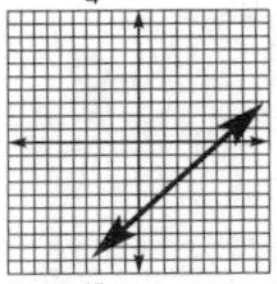

5. (-1, -4), $-\frac{1}{4}$
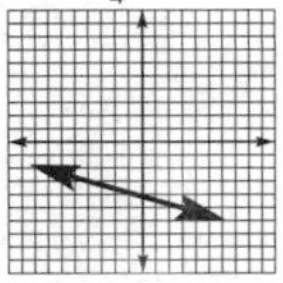

2. (4, 2), 3 (note: $3 = \frac{3}{1}$)
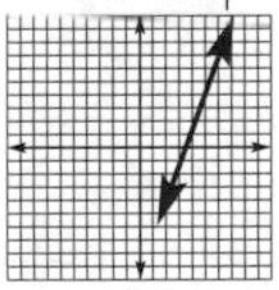

4. (0, 2), $-\frac{5}{2}$
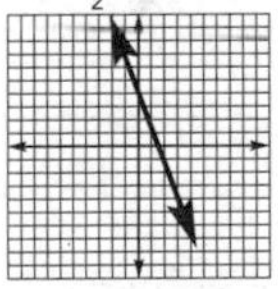

6. (3, -2), 2
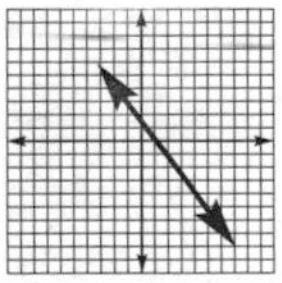

Page 73

1. $y = \frac{1}{2}x - 1$
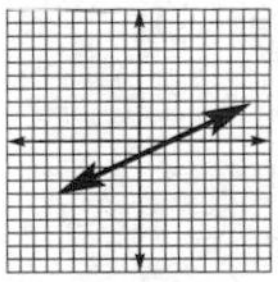

3. $y = -\frac{1}{3}x + 2$
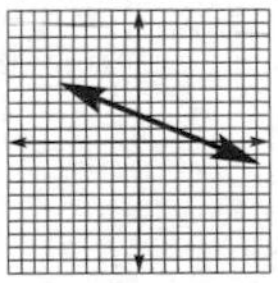

5. $y = -\frac{2}{3}x + 4$
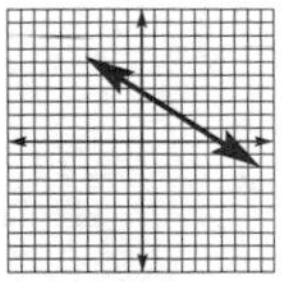

2. $y = 2x + 5$
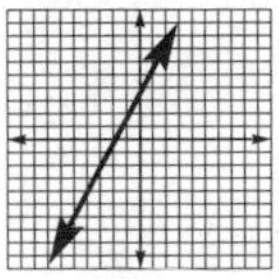

4. $y = -3x - 1$
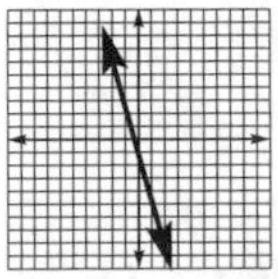

6. $y = x + 3$
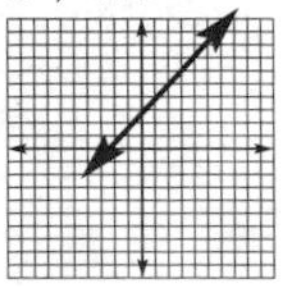

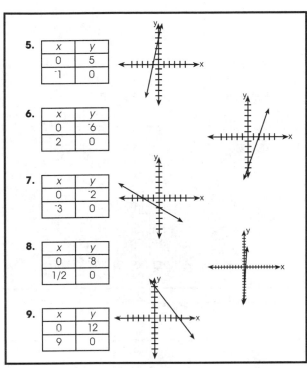

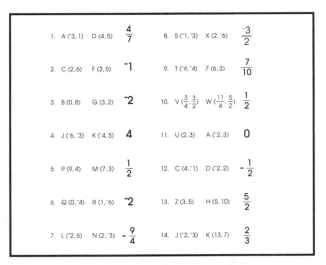

Page 74

Answer Key

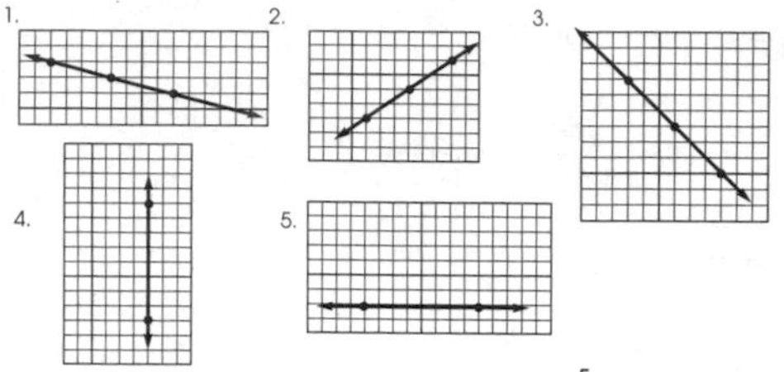

1. $-\frac{1}{4}$
2. $\frac{2}{3}$
3. $^{-}1$
4. no slope
5. 0 slope

6.	(0, 0) (3, 5)	$\frac{5}{3}$	11.	(7, 3) ($^{-}$8, 3)	0 slope
7.	(5, $^{-}$2) ($^{-}$7, 4)	$-\frac{1}{2}$	12.	(0, 0) (4, $^{-}$3)	$-\frac{3}{4}$
8.	($^{-}$6, 3) ($^{-}$2, $^{-}$9)	$^{-}$3	13.	($^{-}$2, $^{-}$3) (2, 5)	$\frac{2}{1}$
9.	(6, $^{-}$9) ($^{-}$4, 3)	$-\frac{6}{5}$	14.	($^{-}$4, 8) ($^{-}$4, $^{-}$3)	no slope
10.	($^{-}$3, $^{-}$11) (2, $^{-}$7)	$\frac{4}{5}$			

Page 75

1. $x + y = 3$ $\quad y = -x + 3$
2. $2x - y = 7$ $\quad y = 2x - 7$
3. $^{-}6 + 2y = 10x$ $\quad y = 5x + 3$
4. $3y - 6x + 12 = 0$ $\quad y = 2x - 4$
5. $2y - 6x = 2$ $\quad y = 3x + 1, m = \frac{3}{1}, y_0 = 1$
6. $y - 4x = {}^{-}3$ $\quad y = 4x - 3, m = \frac{4}{1}, y_0 = {}^{-}3$
7. $4y = 5x + 12$ $\quad y = \frac{5}{4}x + 3, m = \frac{5}{4}, y_0 = 3$
8. $2x - 3y = 5$ $\quad y = \frac{2}{3}x - \frac{5}{3}, m = \frac{2}{3}, y_0 = -\frac{5}{3}$

9. $4x + y = {}^{-}8$
$m = -\frac{4}{1}$
$y_0 = {}^{-}8$

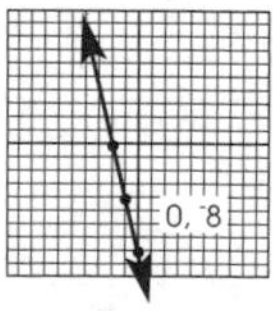

11. $2x - 4y = {}^{-}16$
$m = \frac{1}{2}$
$y_0 = 4$

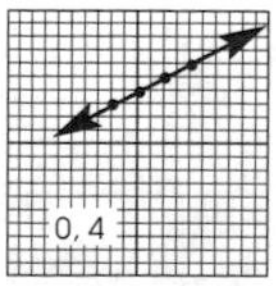

10. $y - 3x = {}^{-}9$
$m = \frac{3}{1}$
$y_0 = {}^{-}9$

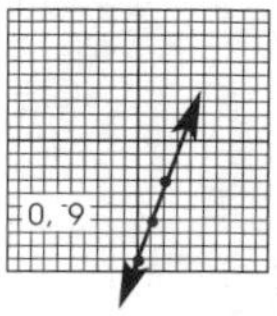

12. $3x + 3y + 4 = 0$
$m = -\frac{1}{1}$
$y_0 = -\frac{4}{3}$

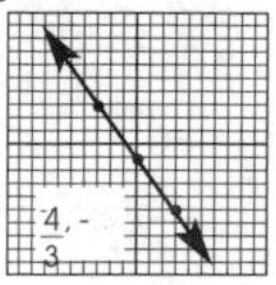

Page 76

1. $y = 2x - 4$

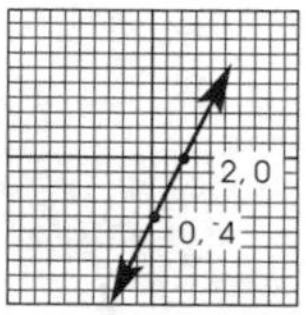

2. $3x - y = 7$

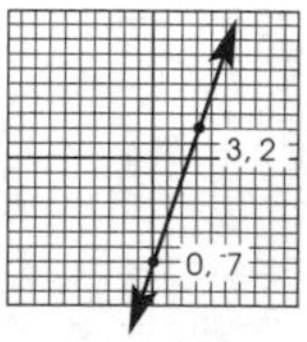

3. $2x + 3y = 6$

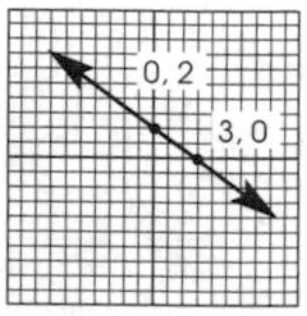

4. $y = -\frac{2}{3}x + 1$

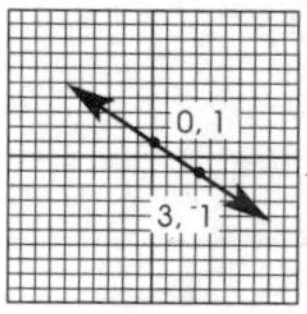

5. $x - 4y + 8 = 0$

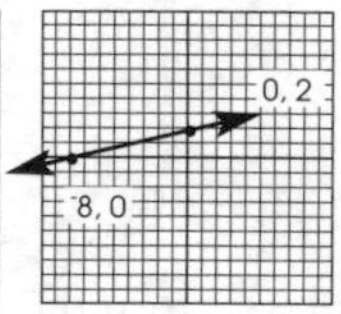

6. $6x - 5y = 15$

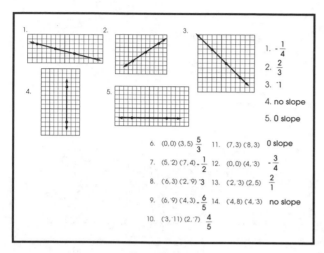

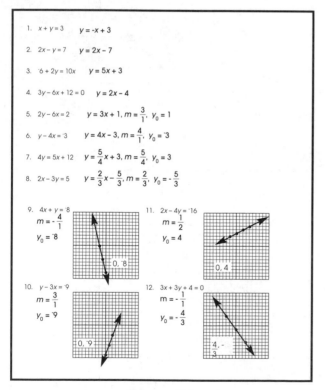

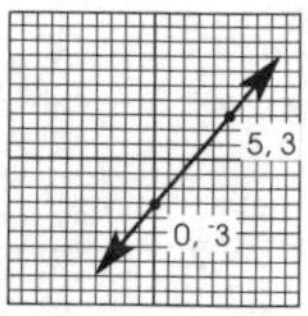

Page 77

1. $y = -\frac{3}{4}x + 2$ $\quad 3x + 4y = 8$
2. $y = \frac{1}{2}x - 2$ $\quad x - 2y = 4$
3. $y = 3x + 6$ $\quad 3x - y = {}^{-}6$
4. $y = -x - 5$ $\quad x + y = {}^{-}5$
5. $y = \frac{3}{4}x + \frac{1}{2}$ $\quad 3x - 4y = {}^{-}2$
6. $y = -\frac{1}{4}x + 8$ $\quad x + 4y = 32$

7. $m = 3$ $\quad y_0 = -\frac{1}{2}$ $\quad 6x - 2y = 1$
8. $m = \frac{5}{4}$ $\quad y_0 = 2$ $\quad 5x - 4y = {}^{-}8$
9. $m = -\frac{2}{3}$ $\quad y_0 = \frac{3}{5}$ $\quad 10x + 15y = 9$
10. $m = 4$ $\quad y_0 = {}^{-}3$ $\quad 4x - y = 3$
11. $m = \frac{3}{4}$ $\quad y_0 = \frac{1}{2}$ $\quad 3x - 4y = {}^{-}2$
12. $m = \frac{7}{2}$ $\quad y_0 = -\frac{3}{4}$ $\quad 14x - 4y = 3$
13. $m = 0$ $\quad y_0 = {}^{-}3$ $\quad y = {}^{-}3$

Page 78

Answer Key

1. $m = {}^{-}3, (4, 5)$
   $3x + y = 17$
2. $m = {}^{-}2, (1, 3)$
   $2x + y = 5$
3. $m = 0, (4, {}^{-}6)$
   $y = {}^{-}6$
4. $m = \frac{3}{4}, (1, 0)$
   $3x - 4y = 3$
5. $m =$ no slope, $({}^{-}3, \frac{3}{4})$
   $x = {}^{-}3$
6. $m = {}^{-}1, ({}^{-}1, 4)$
   $x + y = 3$
7. $m = -\frac{1}{2}, (6, {}^{-}3)$
   $x + 2y = 0$
8. $m = 1, (1, {}^{-}4)$
   $x - y = 5$
9. $m = \frac{1}{4}, ({}^{-}4, 3)$
   $x - 4y = {}^{-}16$
10. $m = \frac{1}{3}, ({}^{-}3, {}^{-}2)$
    $x - 3y = 3$
11. $m = \frac{2}{3}, ({}^{-}1, 1)$
    $2x - 3y = {}^{-}5$
12. $m = 0, (7, {}^{-}4)$
    $y = {}^{-}4$
13. $m = -\frac{2}{1}, ({}^{-}2, {}^{-}7)$
    $2x + y = {}^{-}11$
14. $m = \frac{5}{1}, ({}^{-}2, 0)$
    $5x - y = {}^{-}10$

Page 79

1. $(2, 1)$ $(4, 0)$
   $x + 2y = 4$
2. $(5, 2)$ $(2, {}^{-}1)$
   $x - y = 3$
3. $(4, {}^{-}3)$ $(0, 3)$
   $3x + 2y = 6$
4. $({}^{-}2, {}^{-}3)$ $({}^{-}1, 2)$
   $5x - y = {}^{-}7$
5. $(0, 0)$ $({}^{-}1, {}^{-}2)$
   $2x - y = 0$
6. $(6, {}^{-}3)$ $({}^{-}2, {}^{-}3)$
   $y = {}^{-}3$
7. $(2, 3)$ $({}^{-}1, 5)$
   $2x + 3y = 13$
8. $(4, 8)$ $(4, {}^{-}2)$
   $x = 4$
9. $(5, 8)$ $(3, 2)$
   $3x - y = 7$
10. $({}^{-}2, 5)$ $(3, {}^{-}10)$
    $3x + y = {}^{-}1$
11. $(0, 2)$ $({}^{-}4, 2)$
    $y = 2$
12. $({}^{-}1, {}^{-}1)$ $(0, {}^{-}4)$
    $3x + y = {}^{-}4$
13. $({}^{-}3, 6)$ $({}^{-}3, 2)$
    $x = {}^{-}3$
14. $({}^{-}6, 6)$ $(3, 3)$
    $x + 3y = 12$

Page 80

1. $\frac{{}^{-}2 + 12}{10 - 15} = \frac{10}{{}^{-}5} = {}^{-}2$

2. $\frac{{}^{-}12 - {}^{-}2}{5 - 15} = \frac{{}^{-}10}{{}^{-}10} = 1$

3. $\frac{14 - 28}{{}^{-}3 - {}^{-}1} = \frac{{}^{-}14}{{}^{-}2} = 7$

4. $\frac{6 - 6}{9 - 4} = 0$

5. $\frac{14 - {}^{-}9}{8 - 22} = \frac{23}{{}^{-}14}$

6. $\frac{59 - 0}{33 - 0} = \frac{59}{33}$

7. $\frac{21 - 6}{14 - 14} = \frac{{}^{-}15}{0}$ no slope

8. $\frac{17 - 9}{5 - {}^{-}18} = \frac{8}{23}$

9. $\frac{{}^{-}1 - 9}{16 - 8} = \frac{{}^{-}10}{8} = \frac{{}^{-}5}{4}$

10. $\frac{2 - 9}{{}^{-}4 - 6} = \frac{{}^{-}7}{{}^{-}10} = \frac{7}{10}$

11. $\frac{4 - {}^{-}3}{8 - 7} = \frac{7}{1} = 7$

12. $\frac{4 + 19}{{}^{-}15 + 1} = \frac{23}{{}^{-}14}$

13. $\frac{{}^{-}1 - 6}{{}^{-}8 - 6} = \frac{{}^{-}7}{{}^{-}14} = \frac{1}{2}$

14. #3 & #11; #5 & #12

15. #1 & #13; #4 & #7

16. Every vertical line is perpendicular to every horizontal line.

Page 81

1. slope: ${}^{-}1$, y-intercept: (0,0)
2. slope: 4, y-intercept: (0,0)
3. slope: ${}^{-}2$, y-intercept: (0,0)
4. slope: 0, y-intercept: (0,3)
5. slope: undefined, y-intercept: none
6. slope: 1, y-intercept: (0,1)
7. slope: $-\frac{1}{2}$, y-intercept: $(0, {}^{-}1)$

Page 82

Answer Key

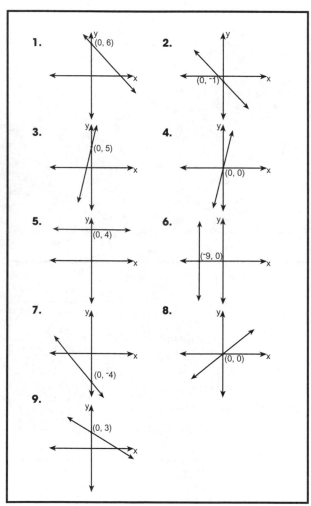

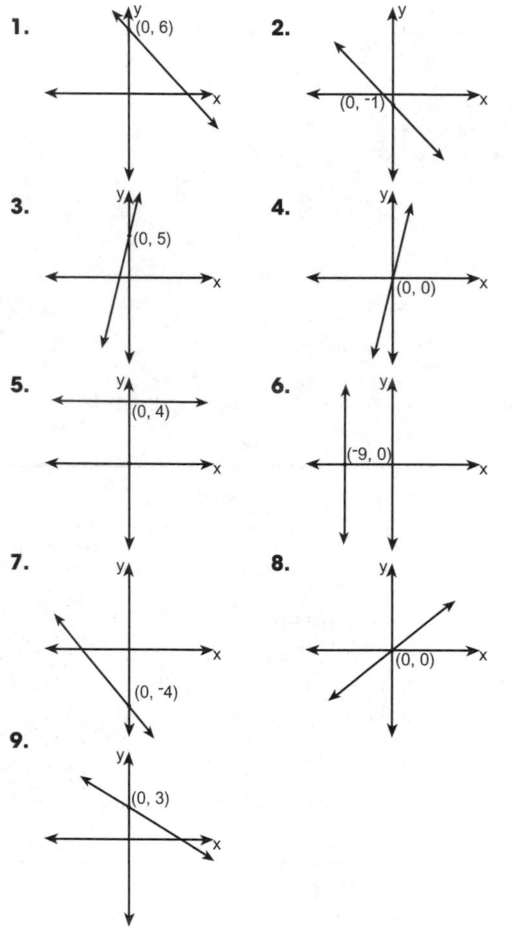

Page 83

1. If it takes 5 minutes to make one cut across a log, how long will it take to cut a 5-foot log into 5 equal pieces? **20 minutes**
2. How can two fathers and two sons divide three automobiles among themselves with each receiving one? **grandfather, father, son**
3. Some months have 30 days, some have 31. How many have 28 days? **all months**
4. If a doctor gave you three pills and told you to take one every half hour, how long would they last? **1 hour**
5. I have two U.S. coins in my hand which total fifty-five cents. One is not a nickel. What are the coins? **half dollar, nickel**
6. Two men are playing chess. They played five games and each man won the same number of games with no ties. How is this possible? **They didn't play each other**
7. Why can't a man living in St. Louis be buried in Illinois? **He is still living.**
8. If dirt weighs 100 lb. per cubic foot, what is the weight of dirt in a hole three feet square by two feet deep? **nothing (no dirt in a hole)**
9. If you had only one match and entered a dark room in which there was a kerosene lamp, an oil burner, and a wood burning stove, which would you light first? **the match**
10. Is there a fourth of July in England? **yes**

Page 84

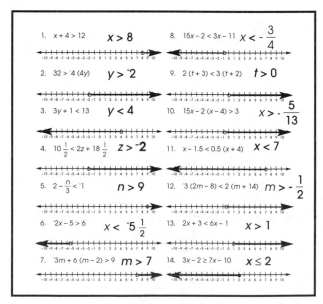

1. $x + 4 > 12$ $\quad x > 8$
2. $32 > {}^{-}4(4y)$ $\quad y > {}^{-}2$
3. $3y + 1 < 13$ $\quad y < 4$
4. $10\frac{1}{2} < 2z + 18\frac{1}{2}$ $\quad z > {}^{-}2$
5. $2 - \frac{n}{3} < {}^{-}1$ $\quad n > 9$
6. ${}^{-}2x - 5 > 6$ $\quad x < {}^{-}5\frac{1}{2}$
7. ${}^{-}3m + 6(m - 2) > 9$ $\quad m > 7$
8. $15x - 2 < 3x - 11$ $\quad x < -\frac{3}{4}$
9. $2(t + 3) < 3(t + 2)$ $\quad t > 0$
10. $15x - 2(x - 4) > 3$ $\quad x > -\frac{5}{13}$
11. $x - 1.5 < 0.5(x + 4)$ $\quad x < 7$
12. ${}^{-}3(2m - 8) < 2(m + 14)$ $\quad m > -\frac{1}{2}$
13. $2x + 3 < 6x - 1$ $\quad x > 1$
14. $3x - 2 \geq 7x - 10$ $\quad x \leq 2$

Page 85

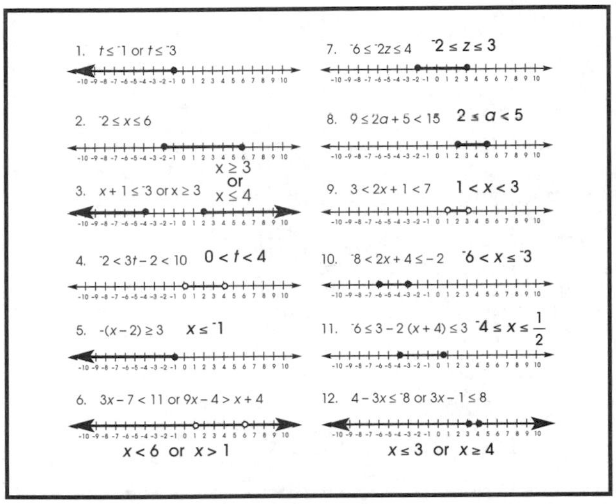

Page 86

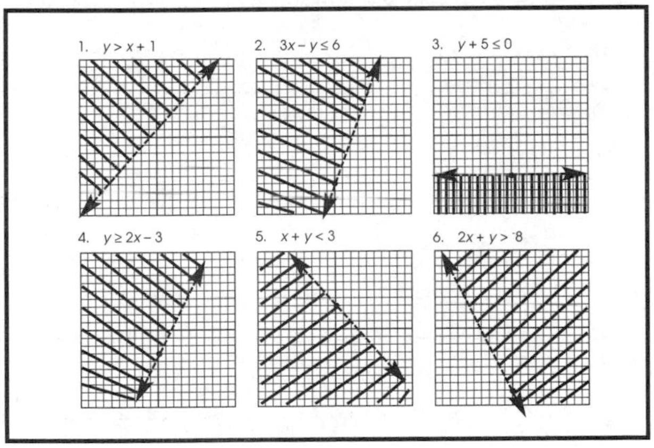

Page 87

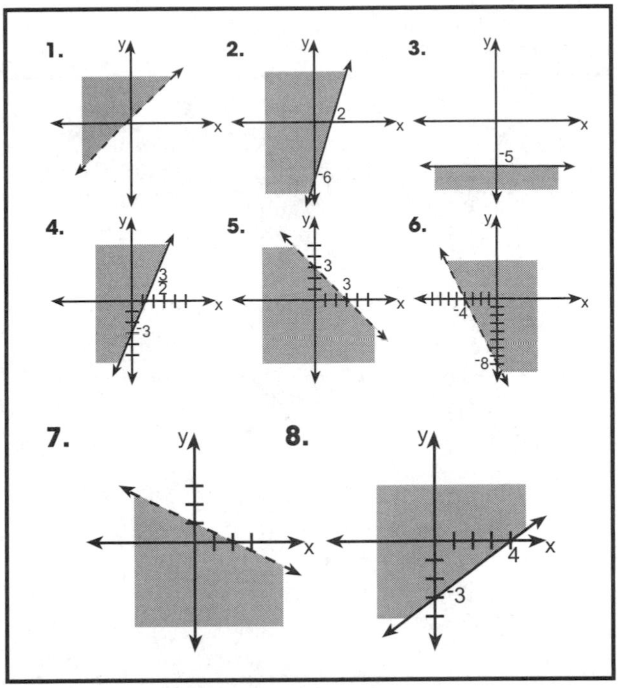

Page 88

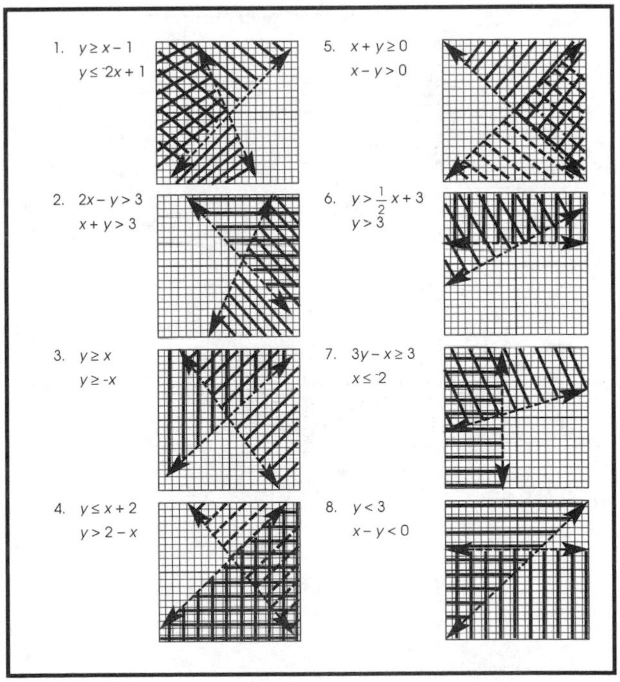

Page 89

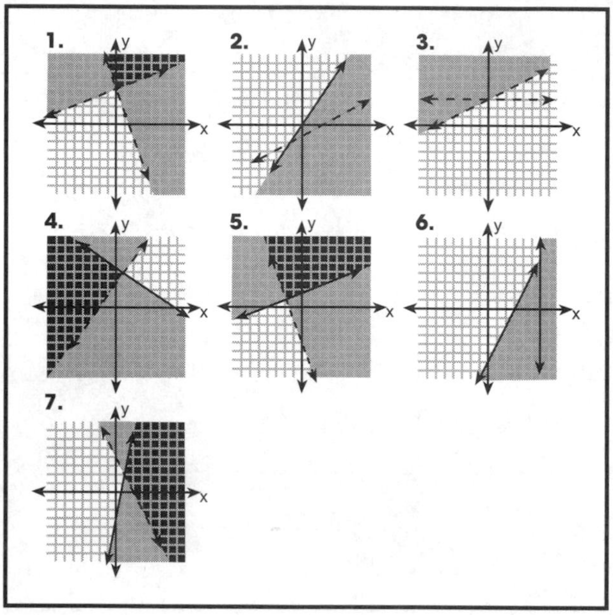

Page 90

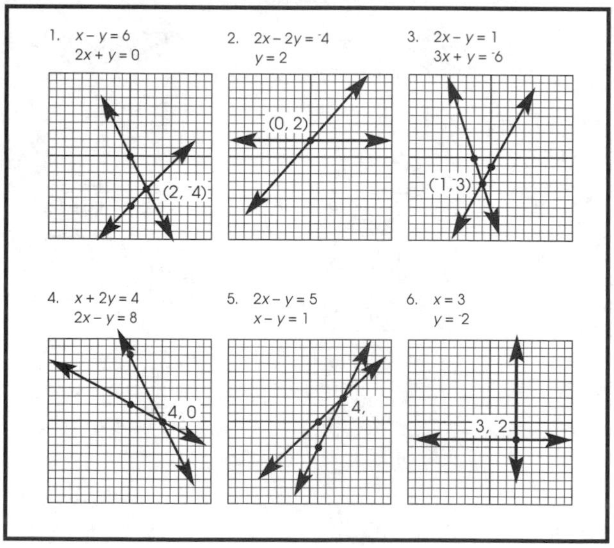

Page 92

Page 91

1. $x - y = 6$
   $2x + y = 0$

2. $2x - 2y = ^-4$
   $y = 2$

3. $2x - y = 1$
   $3x + y = ^-6$

4. $x + 2y = 4$
   $2x - y = 8$

5. $2x - y = 5$
   $x - y = 1$

6. $x = 3$
   $y = ^-2$

1. $2x + y = ^-6$ $3x + y = ^-10$	$(^-4, 2)$	8. $7y + 15 = 3x$ $15 = 3x + 2y$	$(5, 0)$
2. $8x - y = 20$ $^-5x + y = ^-8$	$(4, 12)$	9. $25x = 91 - 16y$ $16y = 64 - 16x$	$(3, 1)$
3. $2x + y = 0$ $2x - 3y = ^-8$	$(^-1, 2)$	10. $4x - 2y = ^-2$ $4x + 3y = ^-12$	$(^-\frac{3}{2}, ^-2)$
4. $5x + 3y = 10$ $2x - 3y = 4$	$(2, 0)$	11. $2x + y = ^-7$ $y = 3x + 3$	$(^-2, ^-3)$
5. $9x - 3y = 9$ $x + 3y = 11$	$(2, 3)$	12. $3x = ^-2y + 10$ $x = 2y + 6$	$(4, ^-1)$
6. $x + 3y = 9$ $x - 2y = ^-6$	$(0, 3)$	13. $x + 4y = 2$ $x - 2y = 8$	$(6, ^-1)$
7. $2x + y = 4$ $2x + 2y = 2$	$(3, ^-2)$	14. $x + 5y + 11 = 0$ $3x - 5y - 7 = 0$	$(^-1, ^-2)$

Page 93

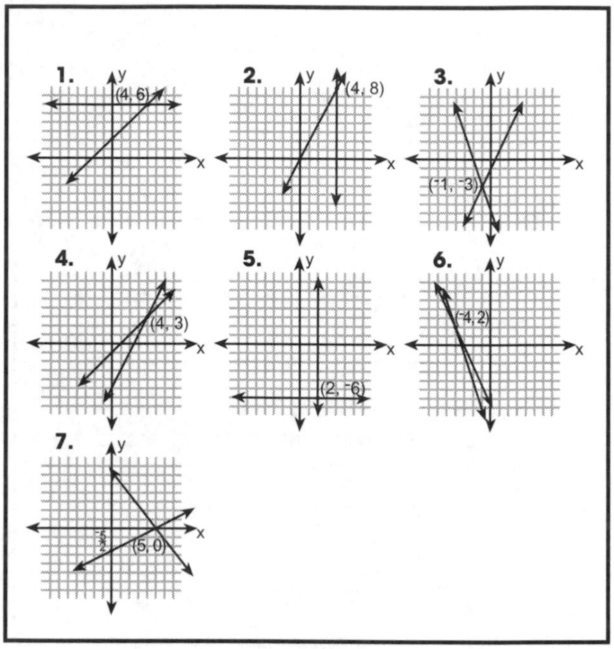

1.
$$4y = 8$$
$$y = 2$$
$$2x + 2 = 0$$
$$2x = {}^-2$$
$$x = {}^-1 \quad ({}^-1, 2)$$

2.
$$\begin{array}{r} {}^-5x + 6y = {}^-16 \\ \underline{5x + y = 2} \\ 7y = {}^-14 \\ y = {}^-2 \end{array} \qquad \begin{array}{r} 5x + {}^-2 = 2 \\ 5x = 4 \\ x = \frac{4}{5} \\ \left(\frac{4}{5}, {}^-2\right) \end{array}$$

3.
$$\begin{array}{r} {}^-4x + 2y = 2 \\ 4x + 3y = {}^-12 \\ \underline{5y = {}^-10} \\ y = {}^-2 \end{array} \qquad \begin{array}{r} 4x + {}^-2({}^-2) = {}^-2 \\ 4x + 4 = {}^-2 \\ 4x = {}^-6 \\ x = {}^-1\frac{1}{2} \\ \left({}^-1\frac{1}{2}, {}^-2\right) \end{array}$$

4.
$$\begin{array}{r} x - 2y = {}^-8 \\ \underline{6x + 2y = 8} \\ 7x = 0 \\ x = 0 \end{array} \qquad \begin{array}{r} 2y = {}^-8 \\ y = 4 \\ (0, 4) \end{array}$$

Page 94

1.	$y = 5 - 4x$ $3x - 2y = 12$	$(2, {}^-3)$	8.	$y = -x + 6$ $x - 2y = {}^-6$	$(2, 4)$
2.	$3x + 2y = 8$ $x = 3y + 10$	$(4, {}^-2)$	9.	$2y - x = 6$ $3y - x = 4$	$({}^-10, {}^-2)$
3.	$3x - 4y = {}^-15$ $5x + y = {}^-2$	$({}^-1, 3)$	10.	$5x - 6y = 16$ $5x + y = 2$	$(\frac{4}{5}, {}^-2)$
4.	$x + y = 2$ $3x + 2y = 5$	$(1, 1)$	11.	$y = 3x$ $x + y = 8$	$(2, 6)$
5.	$x = 3 - 3y$ $4y = x + 11$	$({}^-3, 2)$	12.	$x - 3y = {}^-5$ $2x + y = 11$	$(4, 3)$
6.	$x - y = {}^-15$ $x + y = {}^-5$	$({}^-5, 10)$	13.	$-x + y = 5$ $y = {}^-3x + 1$	$({}^-1, 4)$
7.	$2x + y = {}^-6$ $3x + y = {}^-10$	$({}^-4, 2)$	14.	$2x = 3y$ $x = 3y - 3$	$(3,2)$

Page 95

1.
$$x + 3x = 8$$
$$4x = 8$$
$$x = 2 \qquad y = 3(2) = 6 \qquad (2, 6)$$

2.
$$2(3y - 3) = 3y$$
$$6y - 6 = 3y$$
$$3y = 6$$
$$y = 2 \qquad x = 3(2) - 3 = 6 - 3 = 3 \quad (3, 2)$$

3.
$$x + (x - 2) = 0$$
$$2x - 2 = 0$$
$$2x = 2$$
$$x = 1 \qquad y = 1 - 2 = {}^-1 \qquad (1, {}^-1)$$

4.
$$3x - 2\left(\frac{{}^-5}{4}\right)x = 20$$
$$\frac{3}{2}x + \frac{5}{2}x = 20$$
$$\frac{8}{2}x = \frac{20}{1} \cdot \frac{2}{8} = \frac{40}{8}$$
$$x = 5$$
$$y = \frac{{}^-5}{4}(5) = \frac{{}^-25}{4} \qquad \left(5, \frac{{}^-25}{4}\right)$$

Page 96

1.
$$2x - 4y = 18$$
$${}^-12x + 4y = {}^-88$$
$${}^-10x = {}^-70$$
$$x = 7$$
$$2(7) - 4y = 18$$
$$14 - 4y = 18$$
$${}^-4y = 4$$
$$y = {}^-1$$
$$(7, {}^-1)$$

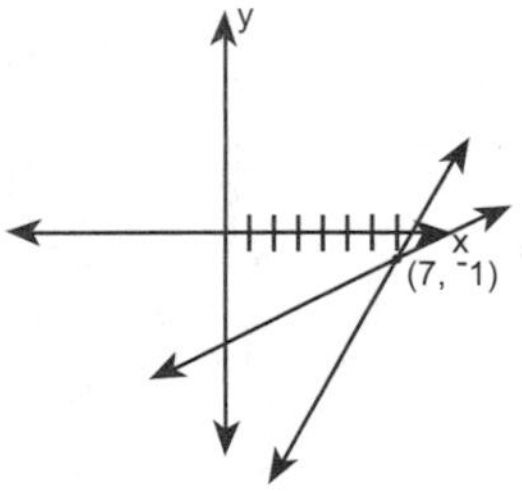

2.
$$3({}^-6y) + 10y = 16$$
$${}^-18y + 10y = 16$$
$${}^-8y = 16$$
$$y = {}^-2$$
$$x = {}^-6({}^-2) = 12$$
$$(12, {}^-2)$$

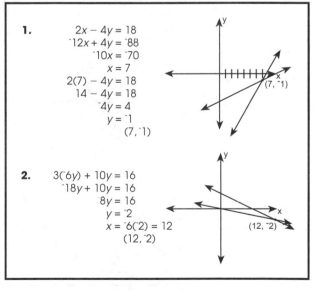

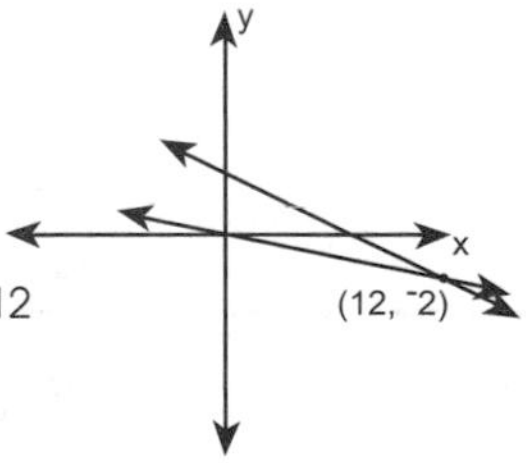

Page 97

3.
$$^{-}3x + 2(5x) = {^{-}28}$$
$$^{-}3x + 10x = {^{-}28}$$
$$7x = {^{-}28}$$
$$x = {^{-}4}$$
$$y = 5(^{-}4) = {^{-}20}$$
$$(^{-}4, {^{-}20})$$

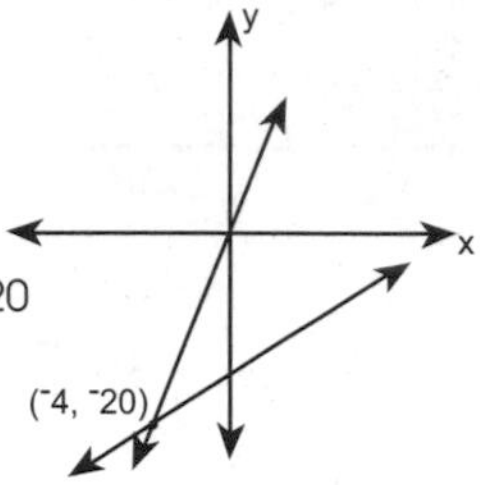

4.
$$28x - 35y = {^{-}133}$$
$$15x + 35y = 90$$
$$43x = 43$$
$$x = {^{-}1}$$
$$^{-}3x + 7y = 18$$
$$7y = 21$$
$$y = 3$$
$$(^{-}1, 3)$$

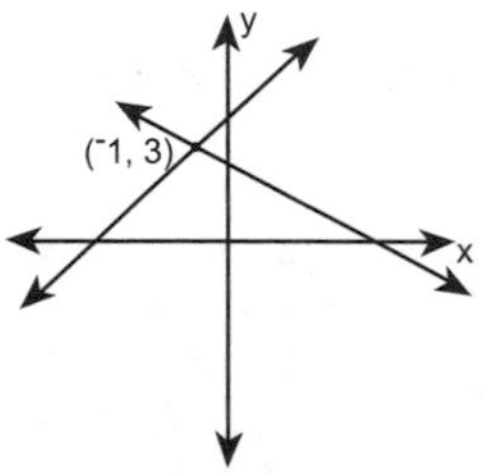

5.
$$5x - 6(4) = 11$$
$$5x = 35$$
$$x = 7$$
$$(7, 4)$$

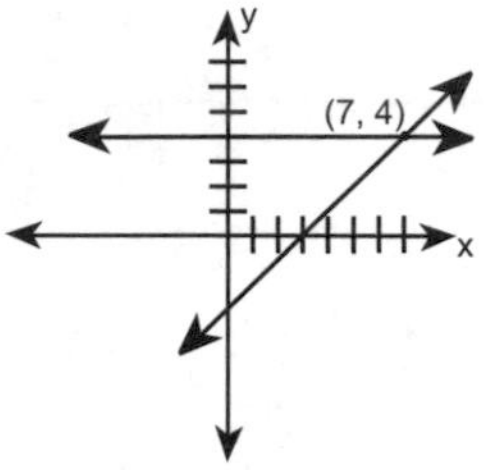

6.
$$2x - (x - 4) = \frac{^{-}5}{2}$$
$$2x - x + 4 = \frac{^{-}5}{2}$$
$$x = \frac{^{-}5}{2} - \frac{8}{2}$$
$$x = \frac{^{-}13}{2}$$
$$y = \frac{^{-}13}{2} - \frac{8}{2} = \frac{^{-}21}{2}$$
$$\left(\frac{^{-}13}{2}, \frac{^{-}21}{2}\right)$$

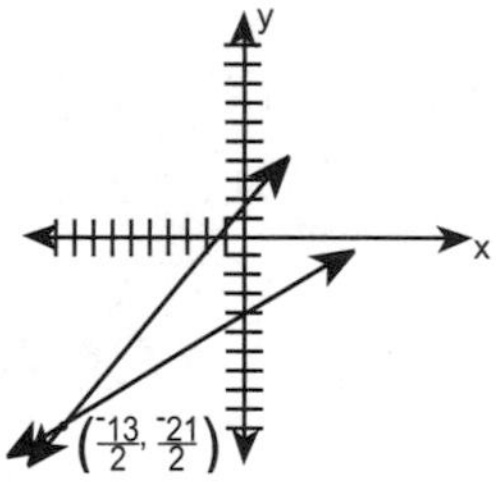

Page 97

1. yes; **2.** yes; **3.** no; **4.** no; **5.** yes
6. yes; **7.** no; **8.** no

Page 98

1. $= {^{-}2}(x^2) + 7 = {^{-}2}x^2 + 7$
2. $= (^{-}2x + 7)^2 = {^{+}4}x^2 - 14x - 14x + 49 = 4x^2 - 28x + 49$
3. $= x^2 - 1$
4. $= {^{-}2}(x - 1) + 7 = {^{-}2}x + 2 + 7 = {^{-}2}x + 9$
5. $= (^{-}2x + 7) - 1 = {^{-}2}x + 6$
6. $= {^{-}2}(x^2 - 2x + 1) + 7 = {^{-}2}x^2 + 4x - 2 + 7 = {^{-}2}x^2 + 4x + 5$
7. $= (4x^2 - 28x + 49) - 1 = 4x^2 - 28x + 48$
8. $= {^{-}2}(^{-}2x + 7) + 7 = {^{+}4}x - 14 + 7 = 4x - 7$

Page 99

1.

x	$f(x)$
$^{-}2$	$^{-}1$
0	3
2	$^{-}1$

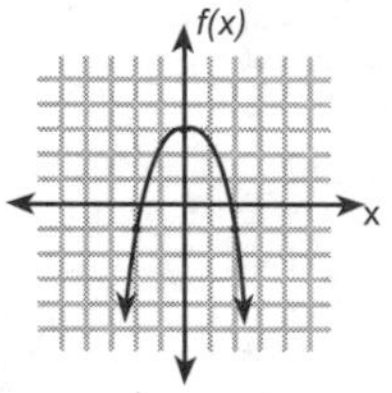

2.

x	$f(x)$
$^{-}2$	2
0	1
2	0

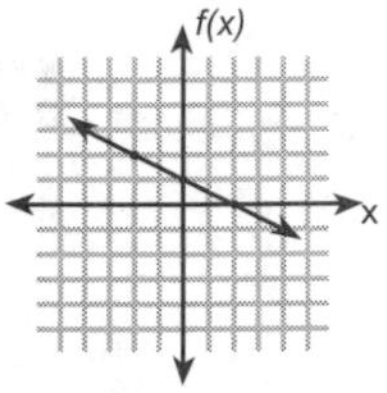

3.

x	$f(x)$
$^{-}1$	4
1	0
3	4

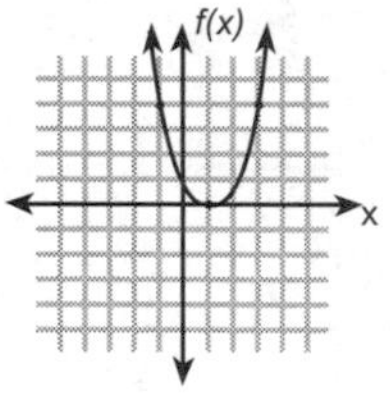

4.

x	$f(x)$
$^{-}1$	$^{-}2$
0	$^{-}4$
1	$^{-}2$

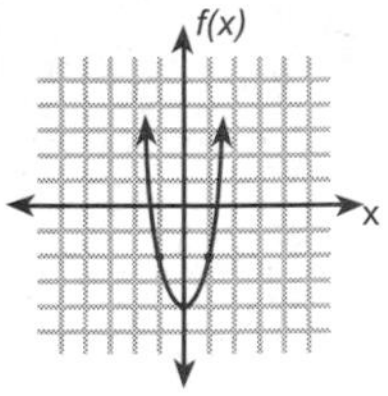

5.

x	$f(x)$
1	$^{-}3$
2	$^{-}1$
3	1

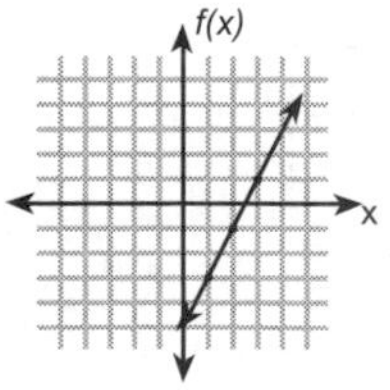

6.

x	$f(x)$
$^{-}2$	2
0	0
2	2

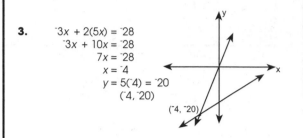

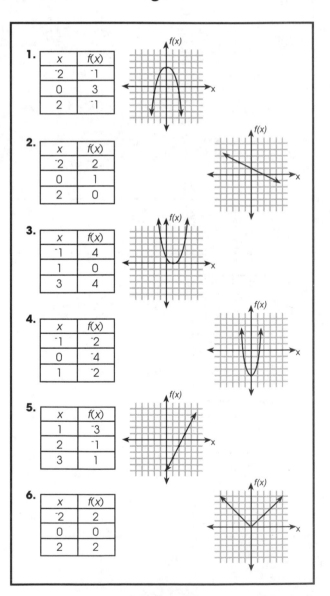

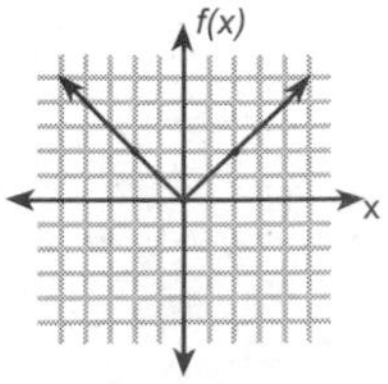

Page 100

Answer Key

7.

x	$f(x)$
1	1
$^-1$	$^-1$
0	0

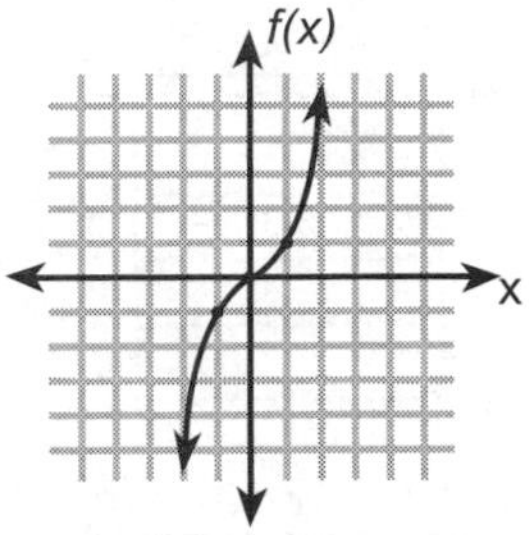

Page 100

1. inverse

x	$f(x)$
2	$^-2$
1	$^-1$
1/2	$^-1/2$
0	0

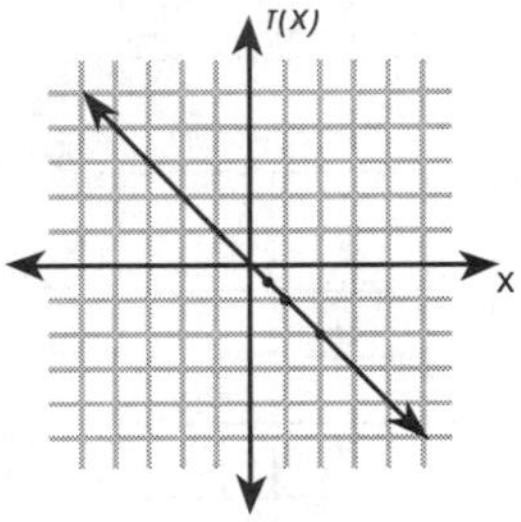

2. inverse

x	$f(x)$
$^-8$	$^-2$
$^-1$	$^-1$
$^-1/8$	$^-1/2$
0	0
1/8	1/2
1	1
8	2

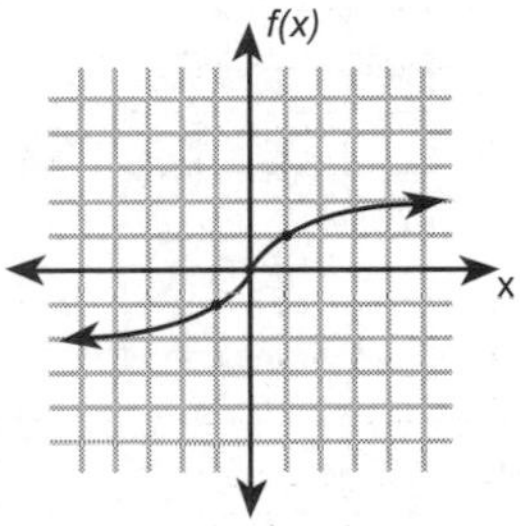

3. inverse

x	$f(x)$
$^-11$	$^-3$
$^-9$	$^-2$
$^-7$	$^-1$
$^-5$	0
1	$^-3$
$^-1$	2
1	3

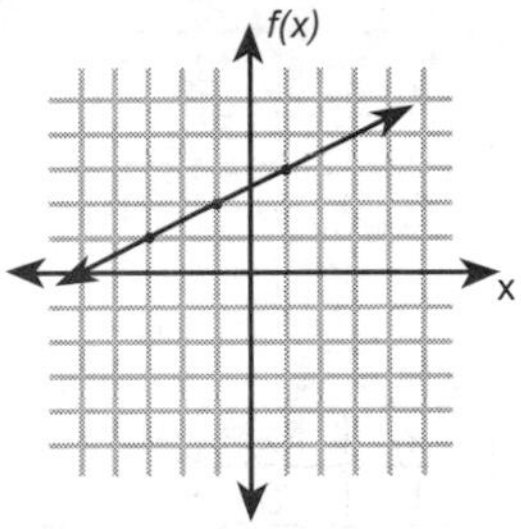

4. inverse

x	$f(x)$
$^-8$	$^-2$
$^-2$	$^-1$
$^-1/2$	$^-1/2$
0	0

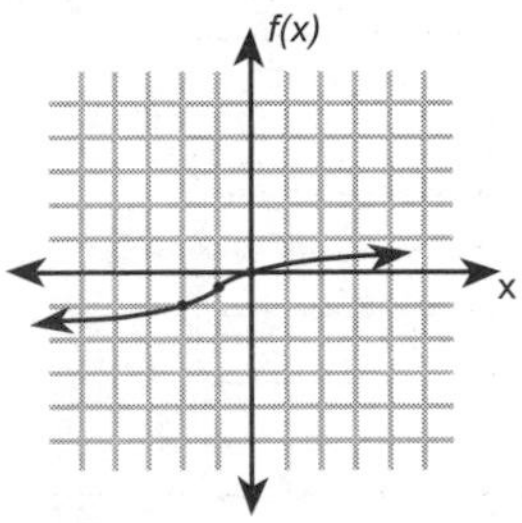

Page 101

5. inverse

x	$f(x)$
8	$^-2$
1	$^-1$
1/8	$^-1/2$
0	0
$^-1/8$	1/2
$^-1$	1
$^-8$	2

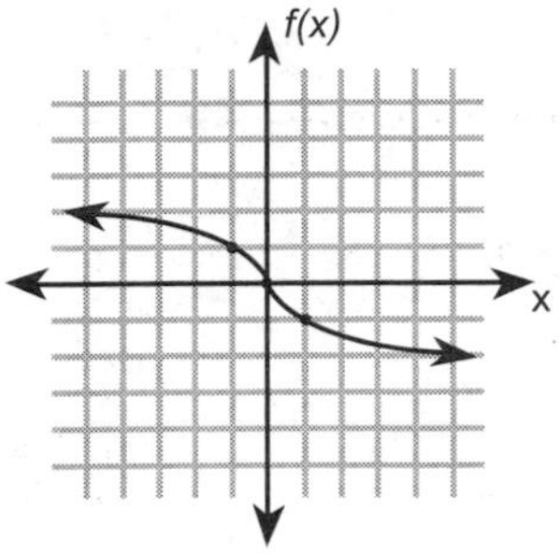

Page 101

1. $x = 2y^2$

$\frac{x}{2} = y^2$

$y = \sqrt{\frac{x}{2}}$

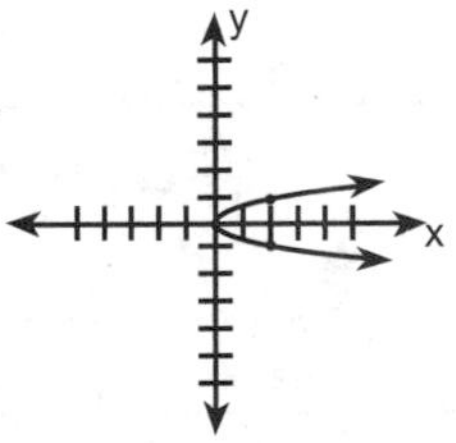

2. $x = y + 2$

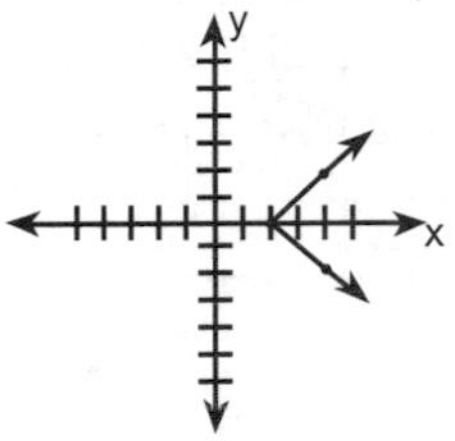

3. $x = -y^2 + 1$

$x - 1 = -y^2$

$-x + 1 = y^2$

$y = \sqrt{-x + 1}$

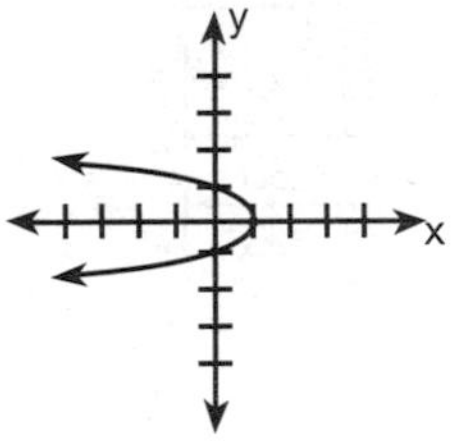

4. $x = \frac{^-1}{2} y^3$

$^-2x = y^3$

$y = \sqrt[3]{^-2x}$

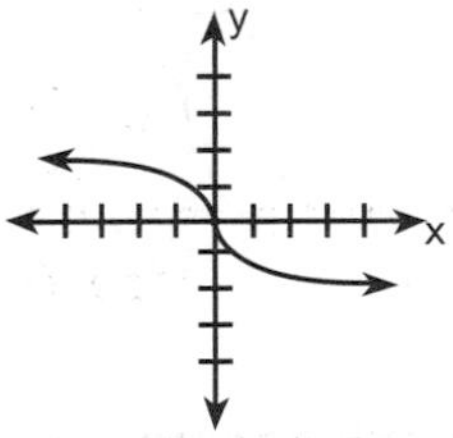

5. $x = y - 1$

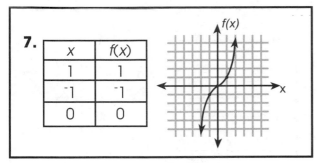

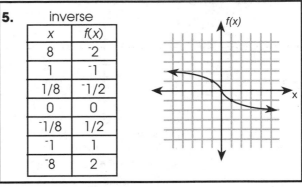

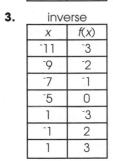

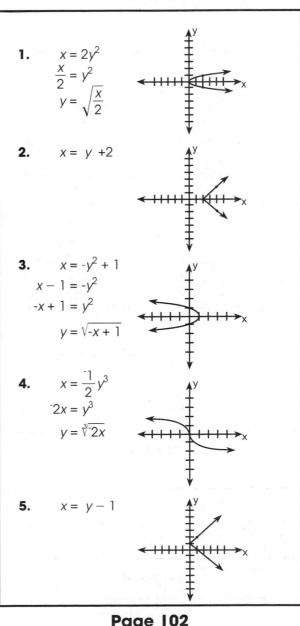

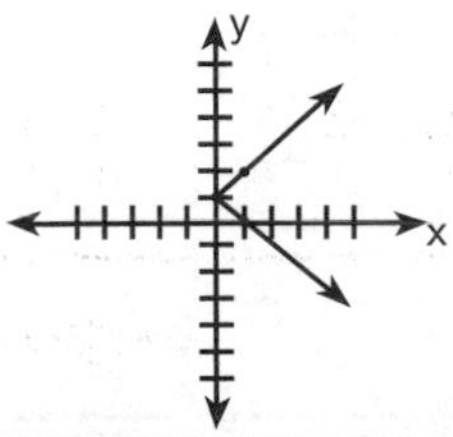

Page 102

Answer Key

6. $x = {}^{-}4y + 7$

$x7 = {}^{-}4y$

$y = x - \frac{7}{^{-}4}$

$y = \frac{^{-}1}{4}x + \frac{7}{4}$

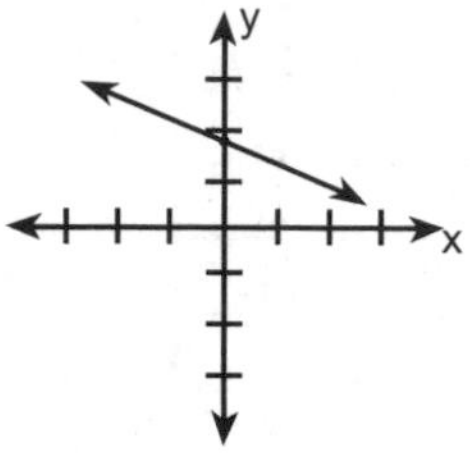

7. $x = -y^2 + 5$

$x - 5 = -y^2$

$-x + 5 = y^2$

$y = -x + 5$

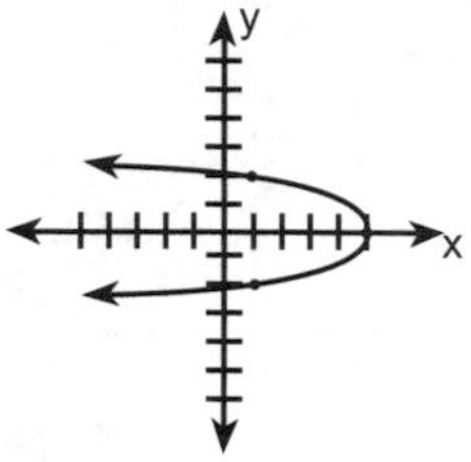

Page 102

1. inverse

x	$f(x)$
$^{-}1$	1/3
0	1
1	3
2	9

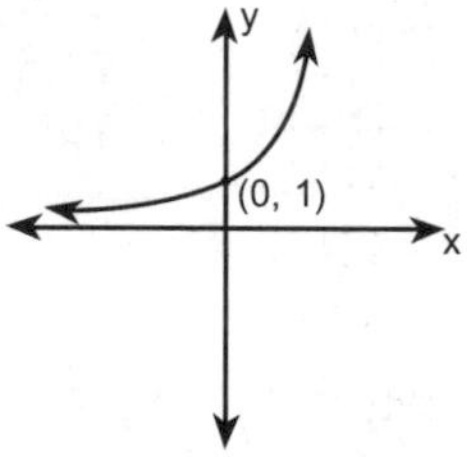

2. inverse

x	$f(x)$
$^{-}1$	3 1/2
0	10
1	30
2	90

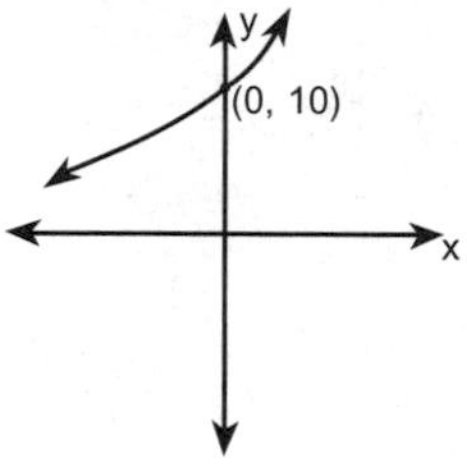

3. inverse

x	$f(x)$
$^{-}1$	5
0	10
1	20
2	40

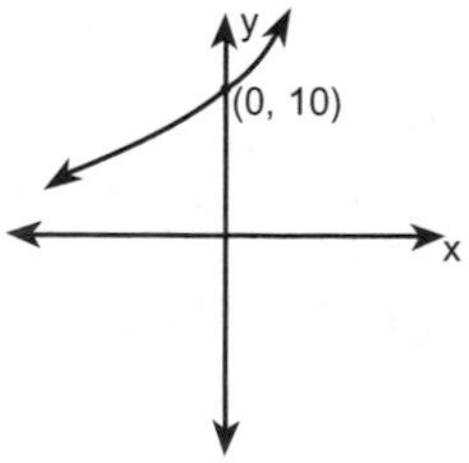

4. inverse

x	$f(x)$
$^{-}1$	2
0	1
1	1/2
2	1/4

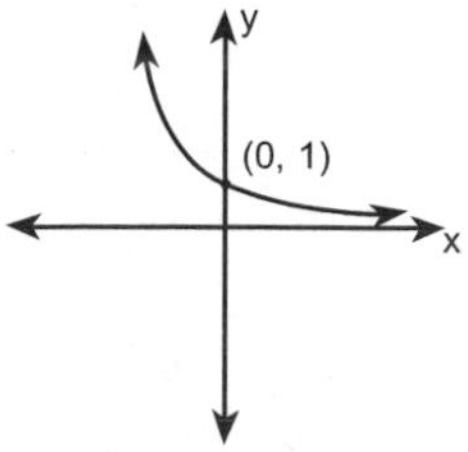

Page 103

5. inverse

x	$f(x)$
$^{-}1$	5
0	1
1	0.2
2	0.04

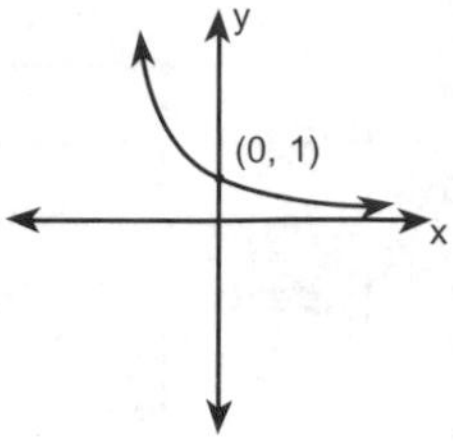

6. inverse

x	$f(x)$
$^{-}1$	12.5
0	50
1	200
2	800

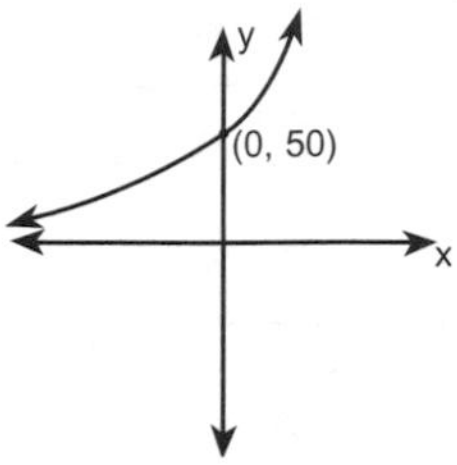

7. inverse

x	$f(x)$
$^{-}1$	0.1
0	1
1	10
2	100

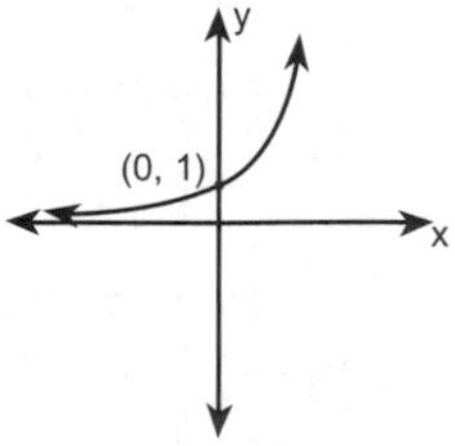

8. inverse

x	$f(x)$
$^{-}1$	10
0	1
1	0.1
2	0.01

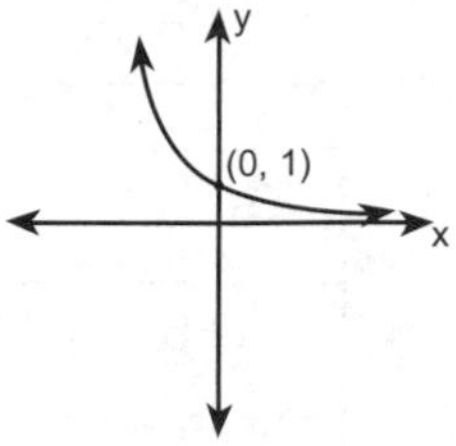

9. inverse

x	$f(x)$
$^{-}1$	25
0	50
1	100
2	200

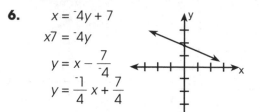

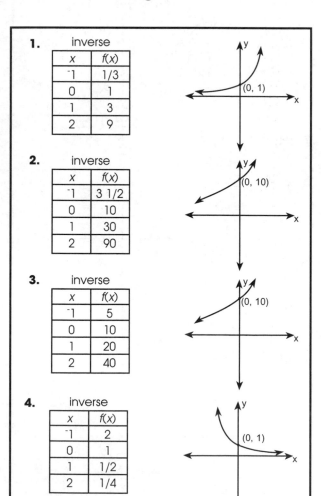

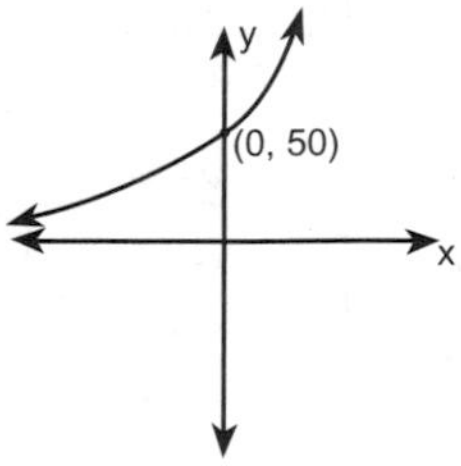

Problems 1, 4, 5, 7, and 8: The y-axis intersects at (0,1). Problems 4, 5, and 8: The *y* value decreases as the *x* value increases.

Page 103